New Frontiers in Regional Science: Asian Perspectives

Volume 16

New Frontiers in Regional Science: Asian Perspectives

This series is a constellation of works by scholars in the field of regional science and in related disciplines specifically focusing on dynamism in Asia.

Asia is the most dynamic part of the world. Japan, Korea, Taiwan, and Singapore experienced rapid and miracle economic growth in the 1970s. Malaysia, Indonesia, and Thailand followed in the 1980s. China, India, and Vietnam are now rising countries in Asia and are even leading the world economy. Due to their rapid economic development and growth, Asian countries continue to face a variety of urgent issues including regional and institutional unbalanced growth, environmental problems, poverty amidst prosperity, an ageing society, the collapse of the bubble economy, and deflation, among others.

Asian countries are diversified as they have their own cultural, historical, and geographical as well as political conditions. Due to this fact, scholars specializing in regional science as an inter- and multi-discipline have taken leading roles in providing mitigating policy proposals based on robust interdisciplinary analysis of multifaceted regional issues and subjects in Asia. This series not only will present unique research results from Asia that are unfamiliar in other parts of the world because of language barriers, but also will publish advanced research results from those regions that have focused on regional and urban issues in Asia from different perspectives.

The series aims to expand the frontiers of regional science through diffusion of intrinsically developed and advanced modern regional science methodologies in Asia and other areas of the world. Readers will be inspired to realize that regional and urban issues in the world are so vast that their established methodologies still have space for development and refinement, and to understand the importance of the interdisciplinary and multidisciplinary approach that is inherent in regional science for analyzing and resolving urgent regional and urban issues in Asia.

Topics under consideration in this series include the theory of social cost and benefit analysis and criteria of public investments, socio-economic vulnerability against disasters, food security and policy, agro-food systems in China, industrial clustering in Asia, comprehensive management of water environment and resources in a river basin, the international trade bloc and food security, migration and labor market in Asia, land policy and local property tax, Information and Communication Technology planning, consumer "shop-around" movements, and regeneration of downtowns, among others.

More information about this series at http://www.springer.com/series/13039

Wardatul Akmam

Arsenic Mitigation in Rural Bangladesh

A Policy-Mix for Supplying Safe Water in
Badly Affected Areas of Meherpur District

 Springer

Wardatul Akmam
Department of Sociology
University of Rajshahi
Rajshahi, Bangladesh

ISSN 2199-5974   ISSN 2199-5982   (electronic)
New Frontiers in Regional Science: Asian Perspectives
ISBN 978-4-431-55153-9   ISBN 978-4-431-55154-6   (eBook)
https://doi.org/10.1007/978-4-431-55154-6

Library of Congress Control Number: 2017951227

Printed on acid-free paper

This Springer imprint is published by Springer Nature
The registered company is Springer Japan KK
The registered company address is: Chiyoda First Bldg. East, 3-8-1 Nishi-Kanda, Chiyoda-ku, Tokyo
101-0065, Japan

To
my parents,
Mrs. Shirin Maqsuda and Professor Dr. Muhammad Shahjahan—
who made it possible for me to be who I am today and set the base for whom I cherish to be in future

Foreword

Wardatul Akmam's book titled *Arsenic Mitigation in Rural Bangladesh: A Policy Mix for Supplying Safe Water in Badly Affected Areas of Meherpur District* has a modest aim at finding a solution to a problem that has put at least 35 million people at risk of arsenic poisoning through drinking water. The book is based on her Ph.D. thesis that was supervised by me. She received the Keiichi Tanaka Award presented by the Japan Section of the Regional Science Association International (JSRSAI) in 2007 for excellence in writing her Ph.D. thesis. Akmam addresses the problem of arsenic contamination in a very comprehensive manner. At first, she reflects on the opinions of experts regarding the cause of contamination of groundwater by arsenic. Next, she explains results of sample surveys depicting the socioeconomic and arsenic-related situation in three arsenic-affected villages in Meherpur district.

The major contribution of Akmam in this book, however, is the model she has developed (proved through computer simulation) in order to supply safe water to the people in a very badly hit village named Taranagar. While developing her mixed-integer model based on Pareto optimality, Akmam has taken into account the exposure to risk of arsenic and bacteria, financial cost of proposed safe water options, walking distance between the source of safe water and the respondents' home, indigenous cultural traits, acceptability of the available options, environmental safety, etc. Furthermore, Akmam has chalked out ways to convince people to change their habits regarding drinking water and proposed a system through which safe water can actually be supplied to the affected people.

Simulation results show that out of a number of available options, the optimal one was dug well + surface water filter. Eighteen dug wells were required in specified zones, so that the distance people had to walk was about 145 meters on an average and no one had to walk more than 360 meters (5-minute distance). For each household, a surface water filter had to be provided to make the dug well water safe from bacteria. Since the first survey carried out in the year 2000, Akmam has carried out some follow-up surveys in the years 2005 and 2011–2012 to monitor the arsenic-related situation in Taranagar. She also provided three safe drinking water sources (dug well/ring wells) for the villagers. In 2015, she has learned from the key informants that at present all the people in Taranagar drink water from safe water sources, although many of them have to spend significant time and energy to procure it. Thus "Taranagar" appears to be a success story in terms of arsenic mitigation.

Emeritus Professor, Faculty of Life   Yoshiro Higano
and Environmental Sciences
University of Tsukuba
Tsukuba, Japan

Preface

Since the 1970s, the people of Bangladesh have been habituated to drink water from tube wells to avoid waterborne diseases like diarrhea, cholera, etc. However, in the late 1990s, tube wells of 61 out of 64 districts of the country were found to be affected by arsenic contamination. The permissible limit of arsenic declared by the Bangladesh authorities is 0.05 mg/l, although the WHO standard is 0.01 mg/l. If one continues to drink arsenic-contaminated water or food for 10 years or more, he or she is likely to suffer from melanosis, leukomelanosis, keratosis, ulcer, gangrene, and cancer of the skin, lung, liver, kidney, and bladder.

Over the last 15 years, many national and international organizations have been and still are working on identifying the cause of the problem, raising awareness among the people, and trying to provide safe water to those who are affected, spending millions of dollars. Nevertheless, in many areas, the actual arsenic-related situation of the affected people has not improved as much as expected. This book in its present form is primarily an outcome of my Ph.D. thesis (completed in 2002) which addressed the socioeconomic aspects of supplying safe water to the arsenic-hit people in three villages in Meherpur district with special emphasis on one village (Taranagar) that was the most affected among the three that were brought under consideration in the beginning. The central focus of the book rests on building a mixed-integer model to find an optimal solution for supplying safe water to the arsenic-affected people in Taranagar. Moreover, ways of convincing the people to drink from arsenic-safe water sources have been discussed, and a system of providing safe water has been developed. An additional chapter has been incorporated in the book, which describes the changes in the socioeconomic and arsenic-related situations in Taranagar village as observed in 2005 and 2012. Thus the book has been written based on information available in different time periods. I request the readers to keep this in mind while they read the book. The model was developed based on information and options of arsenic mitigation available in 2000. However, the model would be useful to find optimal arsenic mitigation option(s) for specific arsenic-affected areas at present and in the future.

It has been reported that 32 million people in Bangladesh still drink water contaminated by arsenic at a higher level than is permissible. The government seems to have forgotten the issue, and donors are showing little interest in it (Daily Prothom Alo, 21 April 2017). Under the circumstances, I believe this book will be of interest to those interested in this field and contribute toward arsenic mitigation in Bangladesh, be it as small as a drop in the ocean. In a work of this kind, there always remains a scope for improvement. So, suggestions from readers are invited and would be accepted gratefully.

Rajshahi, Bangladesh   Wardatul Akmam

Acknowledgments

First of all, I thank Dr. Yoshiro Higano, Professor (Rtd.) and Emeritus Professor of the Faculty of Life and Environmental Sciences at the University of Tsukuba, Japan, who is also the Editor in Chief of *New Frontiers in Regional Science: Asian Perspectives* monograph series, for selecting this proposed book (mostly based on my Ph.D. study) for publication. I am grateful to him for his constructive suggestions and encouragements regarding this publication. I once again express my profound gratitude to Dr. Yoshiro Higano, who supervised my Ph.D. thesis very carefully and sincerely with keen interest, stimulation, and patience, which paved the way for this publication.

I convey my heartfelt thanks to the College Women's Association of Japan for awarding me the Non-Japanese Graduate Scholarship, which partially supported the Ph.D. study during 2001–2002. The CWAJ also awarded me the 60th Anniversary Research Grant in 2009, which allowed me to carry out a follow-up research in the study area during 2011 and 2012. Financial support of the CWAJ has significantly contributed to the completion of the study.

Special thanks go to the geological experts who kindly answered my questions on the cause of the arsenic problem through e-mail communication, despite their time constraints.

My gratitude also goes to my teachers and colleagues in the Department of Sociology, University of Rajshahi, for their constant inspiration regarding my studies in Japan and Bangladesh.

I gratefully thank Dr. Mostain Billah and his associates for helping me in carrying out the interviews among the people residing in Bagoan, Dholmari, and Taranagar in the year 2000. In this regard, I also thank those respondents for their time and opinions. Special thanks go to Mr. Kiamot Ali, a high-school teacher, and Mr. Sagar, a college student residing in Taranagar village, who sincerely helped during data collection and in establishing three sources of safe water (one dug well and two ring wells) in Taranagar village during 2011 and 2012.

My sincere thanks go to Ms. Farida Yeasmin and Ms. Sylvia Mortoza who assisted me in collecting the necessary secondary information in 2000. Sincere thanks also go to former student in the Department of Sociology, University of

Rajshahi, Mr. Md. Golam Martoza and students of the Department of Social Work, University of Rajshahi, Mr. Tushar and Mr. Rakib who sincerely helped in the process of data collection in different phases.

I am grateful to my fellow labmates at the Higano Laboratory of Social Systems Engineering for their help in solving various difficulties I had to encounter being a foreign student in Japan. Thanks from the bottom of my heart go to my Japanese tutor, Dr. Shintaro Kobayashi, and Dr. Rimah Melhem for their continuous help, support, and encouragement. I extend special thanks to my labmates Dr. Takeshi Mizunoya, Dr. Katsuhiro Sakurai, late Dr. Suleyman Ulger, Mr. Kiyonori Taguchi, and Ms. Rie Kurosawa.

I also acknowledge that a few sections of this book have been published earlier as articles in some research journals (e.g., *Studies in Regional Science*, the *Journal of Bangladesh Studies*, and *Papers in Regional Science*). I express my heartfelt gratitude to the editors (Professor Dr. Syed Saad Andaleeb and Professor Dr. Yoshiro Higano) and publishers (including Wiley) who have kindly granted me permission to publish significant portions of those articles again in this book. In this connection, I express my appreciation to the authorities of NGO Forum and especially to Mr. Abul Kalam Raja for providing me some important pictures with permission to publish them in this book.

I am gratefully obliged to my father, Professor (Rtd.) Dr. Muhammad Shahjahan, at the Department of Philosophy, University of Rajshahi, Bangladesh; to my mother, Mrs. Shirin Maqsuda; and to my only sibling (sister), Dr. Farhat Tasnim, for being a constant source of support and inspiration throughout the entire research work. I am particularly obliged to my mother for raising and taking care of my daughter, Faiqua Tahjiba, for two and a half years in Bangladesh in my absence, which enabled me to concentrate more on my studies.

Sincere gratitude is also due to my husband, colleague, and labmate Professor Dr. Md. Fakrul Islam for making all arrangements for my study in Japan and providing me emotional support during the whole research period. I am especially obliged to my only daughter, Faiqua Tahjiba, who was deprived of due maternal care and affection during the whole span of time of this study especially from 1999 to 2003.

Last but not the least, I am grateful to Springer Japan for publishing this book based on empirical longitudinal research.

Department of Sociology   Wardatul Akmam
University of Rajshahi
Rajshahi, Bangladesh

Contents

List of Figures

List of Flow Charts

List of Maps

List of Tables

Chapter 1
Introduction

Abstract This chapter provides a background to this research on arsenic mitigation in Bangladesh with a review of related literature, along with its objectives and methodology. The research addressed the socio-economic aspects of supplying safe water to the arsenic-hit people in three villages in Meherpur district with special emphasis on one village (Taranagar) that was the most affected among the three villages under consideration. Both secondary and primary data were used for this research and the principal methods that were adopted were the social survey and multi-objective mixed integer programming.

1.1 Background

Three decades of relentless efforts made by the Government of Bangladesh and other donor agencies (e.g., UNICEF, UNDP, World Bank) had been successful in achieving the goal of having 90% of the people habituated to drink 'safe' water from tube wells in order to prevent water borne diseases like diarrhea and cholera (Mortoza 2000). However, the so called 'safe' water of many of these tube wells was pronounced 'deadly', owing to contamination of ground water by arsenic above the permissible level in many parts of the country.

Arsenic (As) is a semi-metal odorless and tasteless element, belonging to the nitrogen family. It is naturally found in shale, rock and soil of the earth's crust and in oceans and seas, combined in different forms. Arsenite (As III) and arsenate (As V) are the two forms of inorganic arsenic dissolved in water, the former of which is the most toxic to humans (HVR Arsenic Project 2000).

In 1993, it was discovered that the level of arsenic in the tube well water in Chapai Nawabganj district was far beyond the permissible level. People of the area had developed several symptoms that pointed toward excess of arsenic in their food (and/or drink) intake. In the neighboring province of West Bengal (India), seven districts had already been detected as having tube wells with arsenic contaminated water and people suffering from diseases related to arsenic consumption.

When most of the tube wells in Bangladesh were installed, the drinking water spewed was not tested for arsenic. According to a study carried out in 2000 (DPHE and BGS 2000), tube wells of 53 districts out of 64 had already been affected. The survey mentioned above, discovered that the arsenic content in ground water in

© Springer Japan KK 2017
W. Akmam, *Arsenic Mitigation in Rural Bangladesh*, New Frontiers in Regional Science: Asian Perspectives 16, https://doi.org/10.1007/978-4-431-55154-6_1

different parts of the country varied in between below the detection level (0.25 µg/l) and 1665 µg/l. The worst affected region was the south and southeast region. However, there were some 'hot-spots' in areas, which were known to be generally free of the problem. The DCH and JU study had found that out of its surveyed 60 districts, 41 had tube wells that spewed water contaminated by arsenic above the danger level (DCH and JU 2000).

All the wells located in an area were not affected. It was observed in some cases that water of one tube well was highly contaminated, and the one next to it was safe. Therefore, it was necessary to test the water of all existing (about 11 million) tube wells (Quamruzzaman et al. 1999) in order to ensure that people do not drink arsenic-contaminated water.

The permissible level of arsenic in ground water in Bangladesh is 0.05 mg/l (50 µg/l). The WHO standard was 0.01 mg/l (10 µg/l). Chowdhury T. R. and other authors (Chowdhury et al. 1999a) claimed that up to the level of 22 m, the concentration of arsenic increased with depth, but below that level, the concentration decreased with depth. The DPHE and BGD survey found that the arsenic contamination mainly concentrated within the depth of 150 m. Water of tube wells that were deeper than that level was mostly arsenic free (DPHE and BGS 2000).

The problem of arsenic contamination in ground water in Bangladesh was termed "the largest mass poisoning in history" (Smith et al. 2000), as 80 million people were at risk (DCH 2002). A person who consumes grains produced by irrigated water contaminated with arsenic was also at risk (Ahmed 2000).

Arsenic is a known carcinogen (an element that produces cancer). Someone suffering from diseases caused by excessive arsenic in his/her body is said to be suffering from arsenicosis. A long term (e.g., 5–10 years) exposure to arsenic in drinking water may cause cancer of the lung, liver, skin, kidney and bladder. It also leads to conjunctivitis, hyper-pigmentation, keratosis and gangrene (Mortoza 2000). As we know, in most cases, cancer and gangrene are not curable. Smith et al. referred to the U.S. National Research Council (NRC) to state that exposure to 0.05 mg/l of arsenic "could easily result in a combined cancer risk of 13 in 1000" (Smith et al. 2000).

No specific medical treatment is available for chronic arsenic toxicity. Safe water, multi-vitamins, nutritious food and daily exercise are advised to the patients suffering from arsenocosis (DCH and JU 2000).

In 1997, the government of Bangladesh delegated the official authority of addressing the problem to the Bangladesh Arsenic Mitigation and Water Supply Project (BAMWSP). BAMWSP as a project ended in 2007, before completing quite a few of its objectives. These tasks were transferred to another project named Bangladesh Water Supply Program Project (BWSPP) (see World Bank 2007 and World Bank 2011).

1.2   Outline of the Study

In order to contribute to the efforts in supplying safe water, this study principally focused on developing a theoretical mathematical model in order to find out the guideline through which the people of the study area can make decisions on an optimal safe water option. The study also tested the validity of the model by carrying out simulations. Prior to developing the model, ample information had been collected on the available safe water options and a social survey had been carried out in three highly affected villages in Meherpur, namely, Taranagar, Bagoan and Dholmari to learn about the socio-economic and arsenic related situation of the inhabitants. After arriving at decisive simulation results, the study discussed the mechanisms through which it would be possible to convince the affected people to change their habit regarding drinking water from tube wells and switch to a safe water option. Moreover, the present study chalked out a system through which it might be possible to provide safe water to the affected people. Before making concluding remarks a longitudinal presentation of arsenic related situation in Taranagar Village – the worst arsenic affected village among the three villages under study – has been made.

1.3   Review of Related Literature

There is a vast literature on the magnitude and severity of arsenic contamination in Bangladesh. I have selected only those, which were related to my study. I have divided these literatures into the following categories:

1.3.1. Literature on arsenic problem in Bangladesh in general;
1.3.2. Literature on arsenic-related health risk; and
1.3.3. Literature on arsenic mitigation.

1.3.1   Literature on Arsenic Problem in Bangladesh in General

By 2001 there were two nation-wide surveys carried out on the severity of the arsenic problem (based on sampling). One was by the Department of Public Health Engineering, Bangladesh (DPHE) and the British Geological Survey (BGS) (DPHE and BGS 2001), and the other was by the Dhaka Community Hospital and the School of Environmental Studies (DCH and JU 2000). The latter study was a cumulative one, carried out during 1995–2000. While the former study was restricted to only geological aspects of the problem, trying to locate the most affected areas and the cause behind the problem, the DCH and JU study was also concerned about the people affected – their health related situation. While the former found the problem to be a natural and geological one, the latter surmised that lowering of ground water table has caused the problem in West Bengal (India), and in Bangladesh (DCH and JU 2000).

Hassan and Das explored the spatio-temporal distribution of different levels of arsenic concentrations in Bangladesh, and the impact of this poisoning on health and social life. The paper reports that 70% of the tube wells in the country were spewing water above the permissible level. The authors called for specific intermediate and long-term action to mitigate the problem (Hassan and Das 2000).

Hassan reviewed the approaches in arsenic research in Bangladesh. He focused on the trends in research in this field and identified the gaps to fill in. He emphasized the importance of spatial mapping and generation of arsenic related database on health and social issues (Hassan 2000).

Elizabeth Jones of Water Aid made an overview of the arsenic issue in Bangladesh. She discussed the background and scale of the problem, gave descriptions of 35 organizations active in arsenic mitigation, instrumentation methods for the detection of arsenic, mitigation options, and short term and long-term strategies (Jones 2000).

Melanie Burren carried out a hydrogeological study in Meherpur town and three adjacent villages. This area was quite close to my study area. As such, it is important to discuss the core of this study. From the tube well water samples the author found a wide range of variability in arsenic contamination within a small distance: as low as only 10 m (less than 0.001 mg/l to 0.9 mg/l). Shallow wells showed a wide range of variability while contamination in deeper wells was found to be more consistent, within the middle range. Forty percent of the wells tested within the depths of 10–20 m were contaminated beyond the permissible level. Contamination gradually increasing from that level downwards, became 100% (all the wells were contaminated) at the level of 40 m. The maximum concentration of arsenic was found at shallow levels (about 20 m). Fifty five percent of all the samples examined by the study contained arsenic beyond the .05 mg/l level (Burren 2000).

A paper by WHO proposed a methodology for analyzing the health effects, people's coping strategies regarding the socio-economic consequences of the disease and for making predictions on the beneficial aspects of various mitigation options. It outlined a simulation model and presented an epidemiological sub-model. It made a list of a minimum of core indicators to evaluate the health and socio-economic conditions (WHO 2000).

Fazal et al. summarized the "[e]xtent and severity of groundwater arsenic contamination in Bangladesh", discussing current statistics on the issue citing different sources, and reflecting on the different hypotheses that are being discussed as the cause of arsenic contamination in Bangladesh. The authors stressed proper watershed management and raising awareness among the people (Fazal et al. 2001a, b).

Zuberi evaluated the effectiveness of some of the awareness programs carried out in badly arsenic-affected areas. His observation was that the highest level of awareness that the respondents could reach was five on a seven- point scale. Awareness level of a large proportion of respondents was three or below. He suggested that the materials to which the respondents were exposed were not properly prepared (taking into consideration education level and general beliefs of the people) and the means through which these messages were presented to them was not convincing for them (Zuberi 2003).

Another overall review on the Bangladesh situation regarding arsenic contamination in groundwater was done by M. Ahmed. He reflected on issues such as arsenic and its use, health implications, occurrence in groundwater, acceptable limits of arsenic, possible sources and release mechanism of arsenic in the soil of Bangladesh etc. (Ahmed 2003).

Akmam and Islam identified various factors that influenced awareness level of people on arsenic poisoning, using secondary sources. The identified factors were age, average schooling years of household, exposure to media, number of arsenicosis patients in household etc. (Akmam and Islam 2007a, b).

Findings of a research carried out on nature of awareness in an arsenic-prone village named Khajadanga show that the awareness campaigns to which the villagers were exposed were not at all sufficient for them to achieve a high level of awareness. The highest level of awareness found among the respondents of the village was only five on a scale of nine points (Akmam et al. 2007).

Flanagan et al. referred to the preliminary census statistics collected in 2011 to assume that approximately 19 million people were drinking water with arsenic concentrations >50 μg/l and 5 million people were using water sources supplying water contaminated with arsenic >200 μg/l. The authors believed that if chronic arsenic exposure remains unchecked or worsen in the face of continued population growth and if more private tube wells are installed and used without being having their arsenic levels verified, future generations will be "saddled with enormous health and productivity costs" (Flanagan et al. 2012).

Kevin Krajick and David Funkhouser reported that in Bangladesh, the most hazardous sediments settled down in the last 5000 years. Safer water lied below the 150 m level, within sediments that were more than 12,000 years old. Unfortunately, costs of installing deep tube wells were 10–20 times more as compared to shallow tube wells (Krajick and Funkhouser 2015).

A National Arsenic Policy and Mitigation Action Plan was adopted by the Government of Bangladesh in 2004 to provide arsenic safe water to the exposed population, and to provide medical care to those who had visible symptoms of arsenic related diseases. However, Edmunds et al. regretted that a national monitoring program was yet to be in place. Of the different mitigation strategies that were tested, a large number of small-scale technological remedies proved to be unworkable at the village level. Use of deep groundwater (pumped up from a depth of below 150 m) was the main source of arsenic-safe water for most parts of the arsenic affected regions in Bangladesh. However, in some areas rainwater harvesting was also a means of getting safe water (Edmunds et al. 2015).

1.3.2   Literature on Arsenic-Related Health Risk

The volume of literature concerning health risks associated with drinking arsenic contaminated water is huge. Those considered to be the most important are mentioned below:

In "Human Carcinogenicity of Inorganic Arsenic", Chen et al. found that by drinking water with arsenic 0.01 mg/l/day, the risk of developing skin, lung, bladder, kidney, and liver cancer was 3/1000, 1.2/1000, 1.2/1000, 0.4/1000, and 0.4/1000, respectively, for men, and 2.1/1000, 1.3/1000, 1.7/1000, 0.5/1000, and 0.4/1000, respectively, in case of women (Chen et al. 1997).

According to Smith et al., there was a risk of 1/1000 person dying of arsenic contamination if they continue to drink arsenic contaminated water at the level of .05 mg/l. The authors suggested that the permissible level of arsenic in drinking water should be set at a lower level than .05 mg/l (Smith et al. 1999).

Through simulating data of Samta village of Jessore district, which is very badly affected, Curry et al. predicted the situation after 30 years regarding the arsenicosis situation in the village. More than 22% of the villagers were likely to be patients of arsenicosis and 5.5% of them were likely to die, had the inhabitants of this village continued to drink from the same tube wells (Curry et al. 2000).

Milton et al. tried to relate chronic arsenic poisoning to respiratory effects among Bangladeshis. With an average of 614 µg/l exposure, the overall crude prevalence ratio for chronic bronchitis was 2.3 persons and women were 6 times likely to be suffering from the disease as compared to men. The overall crude prevalence ratio for chronic cough was also 2.9 persons. However, women were at a 10 times higher risk than that of men. As for respiratory distress, the exposed people had twice as much risk of being affected, as compared to those who were not exposed to drinking arsenic contaminated water, and women's risks were 5 times higher than that of men's risks (Milton et al. 2001).

Rahman and Axelson discussed the health effects of arsenic ingestion in Bangladesh. With an exposure level varying from non-detectable to 2040 µg/l, 43.23% of the subjects had skin lesions (average exposure of 7.44 years). Those with skin lesions were 1.6 times more likely to have hypertension. The crude prevalence rate for glucosuria was 1.8 persons (Rahman and Axelson 2001).

In another study it was found that pregnancy outcomes in terms of pre-term birth, stillbirth and spontaneous abortion was significantly higher among those exposed to arsenic in drinking water than those who were not exposed ($p = 0.018$, 0.046 and 0.008 respectively) (Ahmad et al. 2001).

Rice cooked with arsenic contaminated water could be a potential source of arsenic in the bodies of those who eat them. Thus it is an alarming message for the people living in arsenic-prone areas, as rice is the staple (often the only) food eaten by poor people in Bangladesh (Munjoo Bae et al. 2002).

Das et al. have written on the health effects of arsenic contamination in ground water. After giving a historical account of arsenic poisoning, they focused on various health effects of arsenic ingestion into the body – effects on the liver, renal effects, cardiovascular effects, neurological effects, neurologic effects, gastrointestinal effects, respiratory effects, hematologic effects, dermatologic effects, genotoxic effects (mutagenic effects, reproductive effects, carcinogenic effects) etc. The authors also focused on metabolism and excretion of arsenic, along with its diagnosis and treatment. The authors indicated that high protein and vitamin diets helped to clear inorganic arsenic from human body (Das et al. 2003).

Habib et al. examined the relationship between current arsenic levels in ground water and arsenic levels found in urine of the arsenic hit people in order to evaluate the effectiveness of the mitigation programs. The authors reached a conclusion that the green marked tube wells in Bangladesh were not always safe. They recommended that the standard of the WHO was to be followed to remain safe from arsenicosis (Habib et al. 2007).

Brinkel et al. have made a review of literature, which mainly focused on "social consequences and detrimental effects of arsenic toxicity on mental health." Findings of the study show that that "mental health problems (e.g. depression) are more common among the people affected by arsenic contamination." The study points towards "various neurological, mental and social consequences among arsenic affected victims" (Brinkel et al. 2009)   '

1.3.3 Literature on Mitigation of the Problem

In response to the arsenic crisis, the Government of Bangladesh and the World Bank established Bangladesh Arsenic Mitigation and Water Supply Project (BAMWSP) in 1997, which was supposed to be the center (hub) for all arsenic mitigation activities. However, the project could not do much other than a nationwide screening of the tube wells for arsenic and raising awareness (Caldwell et al. 2006; Hanchett and Monju 2009).

A study titled "Arsenic Mitigation in Bangladesh" carried out by UNICEF, Bangladesh focused on the experiences gathered from the UNICEF supported arsenic mitigation programs in different study areas. It reported that people were not very much willing to accept the options of using rain water harvesting and pond sand filter, and still thought that tube well water was safe to drink. The report suggested that the awareness level and change of attitudes was essential to take appropriate actions to establish and maintain the safe water supply sources (UNICEF Bangladesh 1999).

Susan Murcott mentioned various options of safe drinking water, and the relative advantages and disadvantages of different filtering procedures (Murcott 1999). Md. Jakariya (2000) carried out a study titled "The Use of Alternative Safe Water Options to Mitigate the Arsenic Problem in Bangladesh: A Community Perspective". In this study, community response was observed regarding the use of pond sand filter, rainwater harvester, Safi filter, and the three-pitcher filter. He found that people were more interested in drinking deep tube well water, than the options mentioned above. Jakariya also identified some factors that created obstacles towards acceptance of safe water options among the affected people. The main barriers he mentioned included:

1. Not having a clear government policy;
2. Poor economic condition of the people;
3. Preference for the taste of ground water;
4. Peoples' doubts about the effectiveness of the offered options;
5. Lack of coordination among the stakeholders; and
6. Complexity involved in operation of the proposed safe water options.

A study by Bangladesh Rural Advancement Committee BRAC (Ahmed ed. 2000) determined the technical effectiveness of various safe water options and assessed their relative acceptance by villagers in two upazilas (administrative unit) of Bangladesh. Rating the various options in terms of initial and running costs, ease of implementation, requirement for maintenance, supervision, flow rate, suscepti- bility to bacterial contamination, and acceptability to the local community, the study found that the best three options were the three pitcher filter, tube well sand filter and activated alumina filter. All these options involve filtering arsenic from the arse- nic contaminated water.

Report of a Grameen Bank-Department of Health Engineering-UNICEF project revealed the results of an action research on a community based arsenic mitigation project. This study found that on receiving appropriate information, people were able to develop the capability of choosing a proper course of action. In order for the process to continue, it is necessary that the cost be borne by the people themselves. Small, home-based, and less expensive options worked better in this regard, as maintenance (management and maintenance cost) seems to be a problem in community-based options (Grameen Bank 2000).

Kazuyuki Kawahara (2001), a Japan International Cooperation Agency (JICA) expert, has written on the mitigation of arsenic contamination in rural areas of Bangladesh. He emphasized the gradual replacement of short term remedies by long term ones, taking into account the effectiveness of the options (including their capa- bility of sludge disposal, residual chemicals, and meeting water standards), cost of the options (although the initial cost may be paid by outside sources, e.g., the gov- ernment, NGOs, ODA funds, etc. the beneficiaries should be willing to pay for the operation and maintenance), and development of villagers' capabilities with respect to technical management of operation and management of water systems, and col- lection and maintenance of funds. Kawahara also emphasized the mitigation of the medical aspect of the problem, in terms of medical knowledge, cost and develop- ment of villagers' ability to act properly in case of medical necessities (Kawahara 2001).

Yokata et al. discussed the prospects of using pond sand filter (PSF) as an alter- native safe water option in the Samta Village in Jessore, where more than 90% of the tube wells were found contaminated. The authors found that the PSF worked well in Samta and the quality of the treated water was good, if the necessary precondi- tions and precautions are obtained (Yokata et al. 2001).

M. Feroze Ahmed has compiled information on the major potential safe water options that are currently being considered. He discussed the type of areas for which each of the options is suitable, and the preconditions that have to be met for each option as well as their relative merits and demerits (Ahmed 2002).

Mallick and Das have discussed several options for removal of arsenic from ground water, e.g., passive sedimentation, rice husk method, one bucket unit, two bucket unit, three pitcher treatment unit, arsenic removal unit (DPHE-Danida urban water and sanitation project), Steven's technology, Safi filter, activated alumina etc. The authors also discussed the disposal of the wastes produced by these systems, and about some options of bio-solutions and micro-biological processes (Mallick and Das 2003).

Rahman and Jahra have assessed performance of rain-water harvesting as an alternative safe water option in two upazilas of Rajshahi district. Their observations were that rainwater harvesting could provide potentially safe, reliable and afford-able alternative safe water for 8–10 months. However, they also observed that almost 33% of the samples were contaminated by bacteria. The reasons for such contamination was not cleaning the roof tops before collecting water and not fol-lowing the instructions for operation and maintenance properly (Rahman and Jahra 2009).

Performance of nine technologies in providing arsenic-safe water have been assessed by Sutherland et al. Five of these technologies passed the arsenic test com-fortably. The technologies were – Alcan enhanced activated alumina, BUET acti-vated alumina, Sono 3-kolshi, Stevens Institute Technology and Tetrahedron. However, they warned that there was a chance of bacteria-contamination in the fil-tered water and in order to ensure bacteria-free water, regular cleaning with hypo-chlorite and hygiene education was necessary for the users (Sutherland et al. 2009).

The article by Koundouri provides a quick method for cost-benefit analysis of providing safe water to all the people in Bangladesh. According to its findings pond sand filters turned out to be superior to other technologies. However, feasibility of this option to supply water for all areas in Bangladesh is under question. The study shows that with the increase of subsidies for the safe water sources the benefit level decreases. A discount rate of 15% faces a negative net present value (Koundouri 2009).

The main findings of the study by Haque et al. were that higher operation and maintenance cost of community based rainwater harvesting system (CRWHS), arsenic and iron removal plant (AIRP) and Tara pump deterred poor households to adopt these as alternative options; People were willing to pay more to receive better service from deep tube well (DTW), CRWHS and AIRP; Illiterate households pre-ferred DTW and Tara Pump which did not require much maintenance; Media expo-sure affected the decisions of users in selecting alternative options (Haque et al. 2009).

Khan et al. tried to identify the factors that had influence on people's behavior regarding drinking from safe water sources. It was found in this study that current working status and residence (the division in which the respondent resided) were associated with drinking of arsenic-contaminated water (DACW), along with education, ownership of bicycle and television, housing conditions and availability of food all year round. The study recommends that more attention should be given to those who belong to the lower social strata and those residing in Chittagong and Sylhet divisions (Khan et al. 2007).

According to Howard and Ahmed, DTW, RWH, PSF and DW are all viable alternative options. However, DTW offer the lowest disease burden risk and therefore they are the most preferred option. The authors have suggested that dug wells and PSFs should be chlorinated seasonally (Howard and Ahmed 2006).

The study by Paul reveals that awareness of arsenic was still not very much widespread in the study villages, especially those, which were not badly affected. There were gaps in information regarding the consequences of drinking arsenic contaminated water and ways of preventing contamination. The study finds that knowledge of arsenic was associated with education, arsenic-risk region, age and gender (Paul 2004).

Although experts opined that dug well water did not contain arsenic beyond the Bangladesh standard (0.05 mg/l) a study by Uddin and Khan found that 23 out of 51 of their study dug wells contained arsenic more than that level. Five of the dug wells contained water with arsenic exceeding 0.25 mg/l (Uddin and Khan 2004a).

Hoque et al. recommended that arsenic-safe shallow tube wells, deep tube wells and pipeline water systems should be promoted under the existing conditions in Bangladesh. However, the authors opined that a wide range of technologies were yet to be developed and promoted as the options mentioned earlier may not be feasible in all arsenic affected areas (Hoque et al. 2006).

The British Geological Survey urged to rethink abandonment of using deep tube wells as an alternative solution in fear of polluting the deeper aquifers with arsenic. Returning to surface water sources may mean that waterborne diseases will increase. Thus the paper called for a 'pragmatic combination of practical, affordable and sustainable' programs to meet short, long and medium term demands for safe water (BGS 2009).

Inauen et al. (2013) in their research on acceptability of eight alternative arsenic-safe water sources found that the most regularly used options were arsenic removal filters and pipeline water supply. However, piped water supply and deep tube well gained the highest popularity. Those who did not use arsenic-safe source(s) were highly vulnerable, did not like the taste and temperature of the safe water, thought that collecting safe water was too much time-consuming; did not feel the urge to abide by social norms, showed lower level of self-efficacy, and planning to cope with the new situation; and lacked necessary commitment to collect safe water.

Hoque et al., after discussing the pros and cons of the different alternative options, suggested that more research work needs to be carried out on the safe mitigation options. Research on development of suitable options should not only be based on sound science, but also consider local perspectives (Hoque et al. 2003).

Kabir and Howard carried out a survey to see how far the different safe water sources provided to the arsenic hit people were functioning. They found that only 64% of the sources provided were working at the time of the survey. It was observed that community contributions were important in determining whether a water supply source would be functioning or not. The study recommended that improvement of community management is necessary. Caretakers needed more support in terms of technical assistance, machinery and training to repair non-functional sources (Kabir and Howard 2007).

Hussam et al. reported their evaluation on the performance of the SONO filter, which was sufficient for two households, cost around 5000 Taka and was likely to last for at least 5 years. It required simple maintenance. It fulfilled the Bangladesh standard for arsenic and almost cleared all bacteria. The inventors of this filter received the Grainger Challenge Award for this invention (Hussam et al. 2008).

Gadgil and Derby suggested a combination of a mitigation approach based on filtration, ultraviolet disinfection, along with public education. The cost would vary from one place another, but it was estimated that installation cost per person would be around US$10, which was much cheaper than boiling or chlorinizing water (Gadgil and Derby 2003).

A paper by Chen et al. evaluated the impacts of the multifaceted arsenic mitigation efforts to reduce arsenic exposure by interviewing the respondents and measuring levels of arsenic in their urine. They found that arsenic contents reduced by 46% within 2 years time. The reduction had a positive relationship with education, body-mass index, never-smoking etc. (Chen et al. 2007).

A paper by Wagelin et al. evaluated the possibility of accepting SODIS (a solar water disinfection kit to remove bacteria). It required ample supply of high quality pet bottles, which were used to disinfect surface/dug well water. Some people accepted this option while purification of dug well water and purification of water from ponds were not at all acceptable to the people living in the study village (Wagelin et al. 2000).

Lokuge et al. measured the effectiveness of arsenic mitigation programs in terms of disease adjusted life years (DALYs). They found that although the arsenic mitigation efforts had reduced ill effects of arsenic toxicity, it also has a possibility of bringing back the water borne diseases, which haunted the people in Bangladesh in the 1960s and 1970s (Lokuge et al. 2004).

Remmelt et al. (2011) found that arsenicosis was linked to poverty, health and nutrition, to which kitchen gardening could be a remedy. They suggested that long term provisions must be made for medical care, integrated with availability of safe drinking water. These authors also pointed out that the management of the arsenic problem in Bangladesh represented 'a crisis of governance.'

A study by Rahman et al. evaluated the reliability of the arsenic testing field kits. This study observed that 68% of the tube wells that were marked as contaminated were not essentially contaminated and 35% of the tube wells that were marked as safe were contaminated by arsenic. Thus the authors were skeptic about the efficiency of the money used to screen the tube wells in Bangladesh (Rahman et al. 2002).

Milton et al. have evaluated the effectiveness of two arsenic mitigation options – the dug well and the 3-pitcher filter. The study found that although functioning dug wells could be offered as a long term solution its acceptability was only 20% all round the year. Three pitcher filters, unless properly maintained and monitored could not be an effective method for the whole year and might even be harmful in some cases (Milton et al. 2006).

Uddin and Khan evaluated the arsenic contamination in deep tube wells in two unions of Sharsha Upazilla. This paper revealed some new findings regarding arsenic contamination in deep tube well water. It was found that in an area (Dihi) where the contamination rate of normal tube wells was only 11%, contamination rate in deep tube wells was 56%. In another area (Bagachra), 54% of the normal tube wells were contaminated, whereas 40% of the deep tube wells were contaminated. Thus the results of this study challenges the idea that deep tube well water is arsenic-safe (Uddin and Khan 2004).

In the study by Kabir and Johnston performance of different safe water options were observed over one year period. The results showed that majority of the SIDKO plants (78%), ALCAN filters (67%), and SONO filters (58%) were in regular use. However, 15% of the SIDKO plant water, 22% of the ALCAN filter water and 15% of the SONO filter water were contaminated by arsenic well over the Bangladesh standard (Kabir and Johnston 2007).

A study by Sarkar et al. examined how far users in rural Bangladesh were satisfied with their piped water supply systems. The study revealed that almost all the users were satisfied. Eighty six percent of the respondents paid their bills regularly, of whom 50% were well off. Those who were poor reported that they had difficulties in paying the water bills and more than 60% said that they had to reduce consumption of good food as they had to pay a high amount for water (Sarkar et al. 2005).

Ahmed's study revealed the costs of different arsenic safe options to supply drinking/cooking water. The author suggested that although deep tube wells and dug/ring wells were the cheapest options, they were not suitable for all areas. Surface water of acceptable quality was often not available. Cost of household-based rainwater harvesting system with only 50% reliability was very high (Ahmed 2004).

Another study by Jakariya et al. revealed results of a community-based mitigation project involving a population of almost 500,000 people. A total of 403 arsenicosis patients were identified. Active participation of the villagers was ensured in establishing the different safe water options. Different types of community based safe water sources were established, e.g., pond sand filter, rainwater harvesting and different types of arsenic filters. However, these interventions were not long-term solution. The authors suggested that long- term solutions must be based on long term visions (Jakariya et al. 2003).

Ahmad et al. evaluated preferences made by users regarding safe water options in arsenic affected areas. Taking several factors under consideration, e.g., costs, convenience, associated risks and relative advantages and disadvantages, it was found that deep tube well was the most preferred option, followed by three-pitcher filter (Ahmad et al. 2006).

Opar et al. focused on impact of some arsenic-based interventions on 6500 households that included public education, informing people about the state of their tube wells regarding arsenic poisoning and installing 50 community wells. After the intervention, 65% of those who were drinking from unsafe water sources changed their sources, which were often new and untested. It was also observed that distance to the nearest safe water source had an impact on the household responses (Opar et al. 2007).

In his integrated approach, Kunikane included six main aspects of arsenic mitigation, e.g., surveillance of arsenic contamination in groundwater; raising community awareness; community organization; providing a supply of safe water with training to maintain the sources; testing water quality of the sources; and providing health care to arsenicosis patients. Achievements of the project were that they were able to develop 63 'community drinking water supply systems' that were safe and sustainable, community awareness had improved and health conditions of the arsenicosis patients also got better (Kunikane 2009).

Another study by Kamruzzaman and Ahmed assessed performances of the pond sand filters developed by three different institutions – DPHE, NGO Forum, and Danida. Regarding removal of turbidity efficiency of the above were 93.19%, 87.5% and 92.25% respectively. Only 6% of the PSF treated water was coliform free. The study observes that performance of PSF mainly depends on operation and maintenance. One inconvenience regarding use of PSF is that users must depend on another source of water every 1–3 months, when the PSF is cleaned. Expert hand is needed for the cleaning (Kamruzzaman and Ahmed 2006).

Michael and Voss assessed sustainability of deep groundwater as an arsenic-safe water option. Their finding through hydro-geological analysis and simulation covering the whole Bengal basin is that although there is a possibility of downward migration of arsenic by deeper pumping, deeper aquifers may provide a sustainable source of arsenic-free water if its use is kept limited to domestic purposes only (Michael and Voss 2008).

Nahar and Honda, in their review-based article suggested full involvement of community in planning and development of safe water supply systems in their respective localities. In their opinion, community people should take financial and managerial responsibilities and be committed in their endeavor. Continuous monitoring and coordination among the stakeholders were also very important in achieving success in arsenic mitigation (Nahar and Honda 2006).

Barkat and Hussam suggested that Sono Filter should be used for arsenic mitigation to fulfill basic human rights, constitutional rights and justifiable rights of the arsenic affected people in Bangladesh. However, the authors recognized the financial, technical and social constraints involved. The social constraints included myths and misconceptions among people regarding arsenicosis, their lack of awareness and the practice of stigmatizing people with arsenicosis. The authors also observed lack of commitment among the policy planners to boost up the mitigation process (Barkat and Hussam 2008).

A very recent study carried out by Human Rights Watch (2016) has found that 'official response' to arsenic contamination in drinking water, especially in the rural areas is failing. The study also reports "lack of monitoring and quality control". The resources allocated for arsenic mitigation purposes are not being distributed where they are most needed.

1.4   Objectives of the Study

The specific objectives of this research were the following:

1. To understand the nature and severity of the problem;
2. To know the cause of the problem;
3. To carry out a survey among the affected people in the study area to learn about their socio-economic and arsenic related situation, and their views regarding the solution (e.g., willingness to pay for getting safe water, willingness to walk to get water);
4. To gather information on the characteristics (such as cost effectiveness, convenience, etc.) of available options to supply the safe drinking water to the affected people;
5. To develop a theoretical mathematical model and run relevant data through simulation to find out the guidelines for the inhabitants of the study area on how they can choose an option by compromising between the amount of money to be paid, the walking distance they have to cover and the risk they would remain exposed to;
6. To come to a decision regarding the optimal and acceptable solution to the problem of supplying safe water in the study area;
7. To find out the mechanisms through which it would be possible to convince the people to change their habit of drinking water from tube wells and to switch to an optimal safe water option; and
8. To chalk out a system through which it would be possible to provide safe water to the affected people.
9. To portray the differences in arsenic related situations in Taranagar in 2000, 2005 and 2011–12.

1.5   Methodology

Both secondary and primary data were used for this research, and the principal methods that have been used were the social survey and multi-objective mixed integer programming. The author used the census data and findings of other studies that have been carried out so far as secondary information.

In this study, the researcher used the social survey as a method in which data are systematically collected from a sample of population through some form of direct solicitation. This method was used for two purposes. First, an e-mail survey was carried out to learn about experts' opinions on the cause of arsenic contamination in ground water of Bangladesh, and how the problem could be solved. A structured questionnaire was sent to experts in this field. The list of these experts and their e-mail addresses were collected through different website links.

Secondly, a social survey was carried out in December, 2000, among the inhabitants of the three villages under study (which were reported by the local DPHE authorities as the worst affected in that region) to know about their socio-economic and arsenic related situation. I carried out sample surveys among the heads and 'female heads' of households of these villages in order to find out the socio-economic and arsenic related situation. The respondents were selected through simple random sampling. There were about 830 households in Bagoan Village, 437 households in Taranagar Village and 110 households in Dholmari Village. I took a sample of at least 20% of households from each village, and therefore there were 45 respondents from Dholmari, 95 from Taranagar, and 166 from Bagoan. SPSS software has been used to analyze the data. I used some descriptive statistical procedures (e.g., frequency tables, cross tables, averages) and also the χ^2 (Chi-square) test and correlation coefficients.

The heads of households and the 'female' heads of households who participated in this study were interviewed through face to face solicitation. The interviews were carried out by four university students who were familiar to the study area. I myself also visited the villages and talked to many men and women belonging to arsenic affected households.

The basic unit of analysis was the household. Heads of households were interviewed to know about the socio-economic and arsenic related situation of the household. The 'female' heads of household of each of the selected households were especially interviewed to know about their preferences regarding drinking water sources and about the method of its procurement, because according to the ordinary customs in Bangladesh, it is the women folk who are responsible for collecting the water for drinking/cooking. The heads of households who were women were interviewed using questions in both the questionnaires (designed for 'heads of household' and 'female heads of household')

The method I used for developing the model is Multi-objective Mixed-integer Programming based on Pareto optimization. Pareto optimization is a method used to find a specific point at which improvement of any one factor is impossible without worsening others. This optimization had to be multi-objective, as I have taken into consideration, several objectives, mainly, the cost, the distance and the level of exposure to different types of risk. The optimization had to be a mixed-integer one also, as the available options are indivisible in nature.

1.6 Theoretical Background

The model is based on multi-objective Pareto optimization, aiming to show the affected people the information frontier on the basis of which they could select a safe water option suitable for them (through making a trade-off mainly between cost, risk and distance) and thus it is not an endeavor to directly find a suitable decision for any individual, group or authority.

The main idea of Pareto Optimization is to find a solution, which would ensure betterment of one element, without worsening the level of any other element (Liu 2000). As the aim of this research was to identify an optimal guideline, a good use of the idea could be made in order to find the locus of an efficient solution at which no objective function can be improved without worsening at least one of the other objective functions. In other words, instead of seeking a single optimal solution, the target was to find "a set of 'non-dominated' solutions (efficient set, admissible set, Pareto optimal set)" (Zeleny 1974, p. 2).

Research using multi-objective Pareto Optimization began as early as in 1951 by Koopmans (1951) and was followed by Charnes and Cooper (1961), Geoffrion (1967), Geoffrion et al. (1972), and by Philip (1972), among others. In more recent times, the use of this method has increased manifolds. Other approaches of multi-objective programming include utility maximization approach, penalty approach, constraint approach, hierarchical optimization approach, mini-max approach (see for example, Yang 2000), max-mini approach, etc. However, for the purpose of finding a set of optimal solutions by which it would be possible to offer a guideline to the affected people regarding safe water options suitable for them, the author believed that multi-objective Pareto Optimization approach was the best choice. The justification for applying such an approach was that the residents must be informed in advance about the profile of the options by which they would understand that they have to make a compromise and trade-off among several competing objectives on the basis of which they would be requested to make a final decision on which option on the profile they should select. To find such a profile explicitly, the Primitive Pareto Optimization method was rather the most suitable [see for example, "REMARK" on page 113 on the definition given on page 112 in Takayama (1974)].

Another aspect of this model is that the mathematical problem to be solved originally has been formulated as a mixed-integer programming model since most of the available safe water options were indivisible in scale. The indivisibility was essential in the formulation of this model as the policy profile which was aimed to be identified had to give the affected people some useful information when they had to make a decision on what kind(s) of safe water option and how many facilities should be located in which zones in the village by making trade-off between bearing more or less a split installation and maintenance cost of the safe water option(s), walking distance to the closest safe water option, and the risk of being exposed to the amount of arsenic possibly contained in the drinking water provided by the closest one. It is obvious that the most suitable safe water option(s) in the cost and/or risk would be located in every zone with minimum and sufficiently small scale to provide safe water in that zone if the indivisibility had been neglected, namely had we assumed divisibility in scale. Of course, the safe water option of such a small scale would have no reality in most cases and the policy profile based on such an unrealistic assumption would have neglected one of the competing objectives, say, walking distance. Therefore, the mathematical model had to include integer variables.

Integer programming is a kind of linear programming in which the decision variables can only be integers (Greenberg 1971; Sierksma 1996; Walukiewicz 1991). Some of the recent works on and using integer programming include those by Crema (2000), Liu et al. (2000), Alves and Climaco (2000) and Chang (2000). The specific type of integer programming used in the model was "mixed" in the sense that only some (not all) of the decision variables were integers. For this reason the original mathematical problem has been formulated as a mixed-integer multi-objective optimization model. However, the problem which is solved using a mathematical optimization software package, LINGO, is formulated as a (continuous) non-linear multi-objective optimization model, with which we could simulate the step-wise cost function in the integer variable by creating as much nonlinearity as possible in the continuous cost function asymptotically close to the step-wise cost function, and get optimal solutions which would be given by the original mixed-integer one. It would be hard to solve the mixed-integer programming model, which has so many integer variables considering present status of the mixed-integer solution algorithm software package. This is why instead of solving the original mixed-integer model directly, I solved it indirectly by solving the non-linear model.

1.7 The Study Area

The study area of this research comprised of three villages of Mujibnagar[1] Upazila, which were highly affected by arsenic – Taranagar, Bagoan and Dholmari that belonged to the Union (smaller administrative unit under a Upazilla) of Bagoan. These villages were chosen by the researcher, as the Department of Public Health Engineering (Meherpur) informed that these three villages were the worst affected in that Union. Meherpur district had been identified as one of the badly affected districts in the country which contained areas with high concentrations of arsenic (>0.20 mg/l, as shown in the DPHE and BGS (2000) study) and also was a district where 50% to 75% of the tube wells were spewing water containing arsenic above 0.05 mg/l (see Appendix 2, Map 2). However, no significant remedial action was found to be taking place there at the time of the survey. For this reason, I decided to select these villages for my research, which was directed towards finding an appropriate safe water option for the affected people of these villages, as they had been neglected before. Before going on to further details on the specific villages, I intend to describe in brief about Bangladesh as a country.

Bangladesh is a very small deltaic country (area: 144,000 km^2) in South Asia, with a population of about 130 million. More than 90% of the land is covered by alluvial lowland, a gift of the several great river systems, which fall in to the Bay of Bengal (Johnson 1975). Only the districts of Chittagong, Rangamati, Bandarban, Khagrachhari and Sylhet have some hilly areas. The principal river-systems of the country are the Ganges (Padma), Brahmaputra, and Meghna. The land pattern in Bangladesh is changing continually, owing to the changes in the channels and sand banks of the major rivers, and by the floods. The southern part of Bangladesh is mostly occupied by larger part of the Sundarbans, the largest mangrove forest in the world. The climate of the country is tropical, with a highest temperature of 43 °C in

[1] The Union (administrative unit) of Bagoan was located in Mujibnagar Upazilla under Meherpur District. Mujibnagar was separated from Meherpurpur Sadar Upazilla on 24 February 2000.

summer and a lowest of around 3° in winter. Yearly rainfall ranges vary from region to region in between 1500 mm to 3000 mm. November to March is the dry season, during which only about 10% of the rainfall occurs. The rainy season usually continues from June to October, during which 70–80% of the rainfall takes place (Johnson 1975). Arsenic contamination is mostly observed in the alluvial lowland areas, and not in the hilly tracts. However, there has been no research work has found any relationship between arsenic contamination and rainfall and temperature.

Bangladesh is a densely populated country, where the total density of population is 855 people per km^2, with the highest density in Dhaka (5367), and the lowest in the Bandarbans (62) (Government of the Peoples' Republic of Bangladesh 2001b). Relationship between arsenic contamination and differential density is still not known. In hilly areas, (where contamination is low) density is naturally quite low. Dhaka, with the highest density of population is also arsenic free. The economy of Bangladesh is primarily based on agriculture, with paddy, wheat, jute, sugar cane, tobacco and tea, pulses, vegetables and oil seeds and spices as the main crops. The major products of the country include jute, textiles, garments, tea, sugar, paper, matches, leather, cigarette, cement, natural gas, fertilizer, electricity etc. No research to date has investigated the relationship between economic activity and arsenic contamination. The GDP of the country is 187 billion US dollars, and per capita GDP is US$ 1470. About 30% of the GDP comes from agriculture, 17% from industries and 53% from services. Of the total population, 63% are engaged in agriculture, 26% in services and 11% in industries (Virtual Bangladesh 2001).

Meherpur district is situated within the moribund delta region of Bangladesh. The area is less than 9 m above sea level. The only river flowing through the area is the Bhairab, which is almost dead in this region. There is practically no water during the dry season. The region does not experience regular floods; occasional flooding brings about more damage than benefit. Although there are many *bils* (depressions) in the region (Ahmad 1976), dependable sources of surface water supply are rarely found. This is especially true in case of the three villages under the present study. Bagoan and Dholmari are adjacent to Bhairab River, but Taranagar is not. Having the river flowing by the villages of Bagoan and Dholmari is not of much help. Average yearly rainfall in the area is also not very high (around 1500 mm). The villagers use ground water for irrigation. Out of 141,568 acres of cultivable land, 98,413 acres are irrigated (GPRB 2001b). Data received from the Water Development Board of Bangladesh show that although surface water sources become waterless during the dry season, the ground water level does not go below the level of seven meters. Different features of Bangladesh and the study area, such as physical conditions, soil type, rainfall, arsenic contamination, specific area of the study locale are available in Appendix 2.

1.8 Rationale of the Study

From the discussion above, it is very clear that it is necessary to take immediate action in order to supply safe water to the arsenic affected areas. As Bangladesh is a developing country, we must find the optimal safe water option suitable for each of the affected areas. As far as I have gone through, there has not been any study carried out to study the optimality of the safe water option for any area of Bangladesh. Here lies the core importance of this study. Moreover, it is necessary to identify the ways through which it would be possible to convince the affected people to accept the optimal safe water option, and to chalk out the system through which safe water can actually be provided. As none of the above has appeared in research so far, the current research bears great importance. More specifically, the following points depict the importance of this study:

1. It deals with a problem which involves saving millions of lives from being attacked by arsenicosis (disease(s) produced by high amounts of arsenic consumption), and facing an untimely death;
2. The locale of the study has been neglected in terms of providing safe water, despite being very badly affected;
3. The study finds an optimal guideline for the people to choose the most suitable safe water option for them;
4. It takes into account environmental factors in choosing the optimal option;
5. It points out the mechanisms using which the affected people may be convinced to accept the safe water options; and
6. It chalks out a system through which it might be possible to provide safe water to the affected people.
7. Some follow-up surveys have also been carried out to monitor the socio-economic and arsenic related situation in the study village (Taranagar), which makes it a longitudinal study.

1.9 Limitations of the Study

The major limitation of this study was related to availability of data on the different safe water options. While carrying out the optimality study for the safe water options, it would have been better to include some more risk constraints e.g., fulfillment of the standard for manganese, chromium, fluoride, etc., which, like arsenic, are also hazardous for human health if consumed at a level higher than permissible. As I was not able to get such data for all the options, I had to limit my study to include only the risk constraints of arsenic and coliform bacteria. However, the options considered here have been recommended by mitigation authorities as short-term alternative sources of safe water. Even for short-term purposes we must find the optimal and acceptable solution as it is necessary to provide safe water to the people immediately.

Chapter 2
Opinions of Experts on the Cause of Contamination

Abstract In order to find the best possible solution to the arsenic problem, it is necessary to know why and how arsenic dissolves into the ground water in Bangladesh. In this chapter, the author concentrates on the major existing hypotheses on this issue and the results of an e-mail interview carried out among the experts in this field in 2000. Opinions of 12 experts were included. Some scientists believed that the arsenic calamity in Bangladesh was man-made (oxidation of pyrite hypothesis and agrochemical hypothesis), and others argued that the cause of the disaster was rather a geological one (reduction of iron oxyhydroxide hypothesis). Most of the experts (more than 58%) opined in favor of reduction of iron oxyhydroxide hypothesis.

Scientists are not unanimous regarding the cause of the arsenic problem in Bangladesh. According some scientists, the arsenic calamity in Bangladesh is a man-made (oxidation of pyrite hypothesis and agrochemical hypothesis), and others argue that the cause of the disaster is rather a geological one (reduction of iron oxyhydroxide hypothesis). After reading the existing theories, e-mail correspondences with experts in the field have been made, asking them their opinions regarding different aspects of the problem in order to clarify the author's understanding on the issue. These experts were then (in 2000) working in different parts of the world – Bangladesh, India, United Kingdom, the United States, Germany, Australia and Japan. Among the experts, two were Bangladeshi scholar/organization working within the country, and most of the others were Bangladeshi citizens working in other countries. It would have been better to include more local experts, but it was not possible because of lack of e-mail access.

In the replies received, it was found that the experts were quite divided on the issue of the cause of the problem. Before going on to what their opinions were, the gist of the existing hypotheses are discussed.

© Springer Japan KK 2017

W. Akmam, *Arsenic Mitigation in Rural Bangladesh*, New Frontiers in Regional Science: Asian Perspectives 16, https://doi.org/10.1007/978-4-431-55154-6_2

2.1   The Pyrite Oxidation Hypothesis

The first hypothesis to explain the phenomenon of arsenic contamination in ground water in Bangladesh was provided by the "Pyrite Oxidation Hypothesis" (supported by Dr. Dipankar Chakrabarty, Thomas E. Bridge, Meer T. Husain, Del Fanning, Dr. Masud Karim, Dr. Md. Abul Fazal), according to which:

> ...The arsenic contamination results from lowering the water table below deposits of organic matter containing authigenic arsenic pyrites concentrated in organic deposits. These arsenic pyrites oxidize in the vadose (draw down zone) releasing the arsenic as arsenic adsorbed on iron hydroxide. The arsenic adsorbed by the iron hydroxide is released and reduced to its soluble lethal forms when reducing conditions return with subsequent recharge during the rainy season (Bridge and Husain 2000, p. 1).

The supporters of the oxidation theory argue that arsenic in ground water is a recent phenomenon. Excessive extraction of ground water has caused the water table to go down to lower levels, and hence the problem (Chowdhury et al. 1999a). Some of these scholars (Bridge and Husain 2000) argue that by building dams at the up-stream, India withdraws water from rivers that flow through India into Bangladesh during the dry season. This has caused the drying up of rivers within Bangladesh territory, which made the farmers use ground water for irrigation at a massive scale, and this over extraction of ground water has caused the arsenic problem. Therefore, the operation of the barrages built by India is to be blamed for the arsenic problem in Bangladesh.

2.2   The Reduction of Iron Oxyhydroxide Hypothesis

After intensive observation of 46 wells in Bangladesh, R. Nickson et al. (Nickson et al. 1998, p. 338) discovered that their findings point to a different mechanism –

> ...in oxic (shallow) wells, arsenic concentrations are mostly below 50 µg l^{-1}; in anoxic waters, arsenic concentrations (=260 µg l^{-1}) correlate with concentrations of dissolved iron (=29 mg l^{-1}) and bicarbonate...; and arsenic concentrations increase with depth in wells at Manikganj, Faridpur, and Tungipara. These observations suggest that arsenic is released when arsenic-rich iron oxyhydroxides are reduced in anoxic groundwater, a process that solubilizes iron and its absorbed load and increases bicarbonate concentration. Sedimentary iron oxyhydroxides are known to scavenge arsenic and, in Ganges aquifer sediments, concentrations of diagenetically available iron (≤3.7%) and arsenic (≤26 ppm) correlate well.
> The arsenic-rich groundwater is mostly restricted to the alluvial aquifers of the Ganges delta. The source of arsenic-rich iron oxyhydroxides must therefore lie in the Ganges source region upstream of Bangladesh. Weathered base-metal deposits are known to occur in the Ganges basin (at Bihar, Uttar Pradesh, West Bengal), so weathering of these arsenic-rich base-metal sulphides must have supplied arsenic-rich iron oxyhydroxide to downstream Ganges sediments during Late Pleistocene- – Recent times. The arsenic-rich iron oxyhydroxides are now being reduced, causing the present problem. Reduction is driven by concentrations of sedimentary organic matter of up to 6%.

The survey conducted by Department of Public health Engineering (DPHE), Bangladesh, and British Geological Survey (BGS), U.K. (2000) examined 3534 tube wells in Bangladesh. According to this survey also the strongly reducing environment in the sediments of Bangladesh is responsible for arsenic to dissolve into the water.

Fazal et al. (2001a) have tested the validity of the DPHE and BGS (2000) study. Although their results were more or less similar to those found in that study, Fazal et al. discovered errors in the estimates of the percentage of wells contaminated. Moreover, it was detected that "space-dependent relationships among different hydro-geological parameters for the heterogenous aquifer system (were) not valid to represent time-dependent phenomena" (p. 70). Based on this validity study, Fazal et al. concluded that the conclusions made by the DPHE and BGS (2000) study were not valid, and that there was still possibility that the pyrite-oxidation hypothesis would prove to be acceptable.

Some scholars believe that both reduction and oxidation theories may be correct, depending on the nature of the sediments (As-contamination Research Group, Niigata University 2000). So far, there has not been any scientific evidence to prove the oxidation theory. The reduction theory is also not able to answer all the questions perfectly. For example, the reduction theory is unable to explain the existence of the 'hot spots' in a generally low-arsenic area. It is also unable to show why and how the reduction is occurring now, and had not occurred before. There are weak points in both theories. However, both theories deserve further consideration. In relatively recent times authors (Shankar et al. 2014) have expressed their opinions in favor of the reduction of iron oxyhydroxide hypothesesis.

2.3   The Agrochemical Hypothesis

Dr. Jamal Anwar has written a book (Arsenic Poisoning in Bangladesh: End of a Civilization?) (Anwar 2000), in which he points to the uncontrolled and indiscriminate use of sub-standard chemical fertilizers and pesticides (often not allowed to be sold in places of manufacture), in the name of "Green Revolution" for three to four decades in Bangladesh as one of the very important causes of arsenic contamination in ground water. Naturally, the soil of Bangladesh is poor in phosphate. However, wherever high concentrations of arsenic is detected in ground water, phosphate is detected along with it. Dr. Jamal Anwar refers to a study carried out in the province of Uttar Pradesh, India, which discovered 2% arsenic concentration in super phosphate fertilizers. He suspects that the same kind of fertilizer has been used in Bangladesh also, which has given rise to the current arsenic problem. However, this hypothesis is still to be proven scientifically.

Table 2.1 Hypothesis supported by respondents

Hypothesis	Number of respondents	Percentage
Oxidation of Pyrite	4	33.33
Reduction of Iron Oxyhydroxide	7	58.33
Agrochemical	1	8.33
TOTAL	12	100.00

Table 2.2 Response regarding over-extraction of groundwater – arsenic contamination relationship

Response	No. of respondents	Percentage
Yes	9	75
No	3	25
Total	12	100

2.4 Summary of the Responses of Experts

After collecting the names and e-mail addresses of experts from the internet, the author sent a questionnaire to 26 experts and researchers regarding the cause of the arsenic problem in May 2000. Among them, 16 responded. Four of these researchers expressed their inability to express their opinions on these specific questions. Answers from 12 experts were received.

The questionnaire included questions regarding the hypothesis which the experts considered more convincing, relationship between arsenic contamination and the barrages built by India at the upstream of international rivers, whether over-extraction of ground water had any impact on the problem or not, the depth level of arsenic contamination, and suggestions for a safe water supply. Below, the answers are discussed with the help of some tables.

Table 2.1 shows that there is a tendency among the respondents to accept the reduction theory in higher number. However, the reasoning given by the 'oxidation of pyrite' theorists cannot be nullified either. For the agrochemical hypothesis, more scientific proof is needed. Experts who still supported the oxidation theory, said that they were not convinced about the validity of the data presented by the reduction theorists. The reduction theorists on the other hand argued that the actual characteristics of the sediments of Bangladesh soil do not conform to the oxidation theory.

It is interesting to observe in Table 2.2 that although only 4 of the respondents accepted the oxidation theory, 9 of them believed that there was a relationship between over extraction of ground water and the arsenic problem. This gives us the impression that we should reduce using underground water, as much as possible. However, Table 2.3 portrays a different picture regarding respondents' opinion on the issue of relationship between the barrages built by India and the arsenic problem. Although Masud Karim, Md. Abul Fazal argue that there was a relationship between the two variables, Gawher Nayeem Wahra has pointed out that there was no proof regarding this issue, and that the upstream of the Farakka Barrage was also affected by the arsenic problem.

Table 2.3 Response regarding the relationship between barrages (built by India) and the arsenic problem in Bangladesh

Response	No. of respondents	Percentage
Yes	3	25
No	9	75
Total	12	100

Table 2.4 Response regarding depth of arsenic contamination

Response	No. of respondents	Percentage
Increases up to 15–20 m, then decreases	2	16.67
10–30 m	2	16.67
20–50 m	4	33.33
No specific depth level	4	33.33
Total	12	100.00

Table 2.5 Response regarding suitable alternatives for mitigating the problem

Response	No. of respondents	Percentage
Deep tube well/rainwater harvesting/filtering	1	8.33
Purified surface water/rain-water harvesting	2	16.67
Dug (Shallow) wells	2	16.67
No specific answer (any/all of the above)	7	58.33
Total	12	100.00

Table 2.4 portrays the depth level at which the experts think arsenic contamination is the gravest. Four (33.33%) of them did not give any specific answer. Two (16.67%) of the respondent experts thought that contamination increased up to 15–20 m and decreased below that level. Four of them opined for the level of 20–50 m, two (16.67%) believed that the highest contaminated level is 10–30 m deep. So, we see that the scientists were not unanimous regarding the depth level of contamination.

In the e-mail survey, the experts were requested to mention the means of supplying safe water, which they thought was suitable for mitigating the current arsenic problem in Bangladesh.

All the experts and researchers agreed that in order to suggest the most appropriate solution to the problem, identification of the actual cause of the problem through intensive research was essential. In response to the question regarding suggestions for a safe alternative to tube well water, most of the respondent experts refrained from giving any specific answer (see Table 2.5). One of them (8.33%) opted for deep tube well, rain water harvesting and filtering, two (16.67%) suggested dug well, and two (16.67%) were in favor of treatment of surface water and rainwater harvesting.

Therefore, from the e-mail survey, we may conclude that although researchers were not unanimous regarding the cause of arsenic contamination in ground water in Bangladesh, larger number experts were convinced by the reduction hypothesis. Their opinions varied significantly on the depth-level of the contamination, and most of them did not specify the most suitable safe water option. However, 75% of the respondent experts believed that there was a direct or indirect relationship between over-extraction of ground water and arsenic problem in Bangladesh. In an effort to reconcile the opinions of the experts, we may say that there is a greater possibility that over-extraction of ground water has contributed to the arsenic contamination in ground water, either directly, or indirectly. While more intensive and detailed research has to be carried out to be certain about the actual cause of the problem, use of ground water for irrigation has to be discouraged, although this is likely to have grave negative impact on the agricultural production in Bangladesh, as farmers mostly use ground water for irrigation, which has contributed to the increase in total production significantly. More intensive research should be carried out to find the advantages, disadvantages and optimality of the different alternative ways of irrigating that are suitable for Bangladesh.

Chapter 3
Social Survey in the Study Area

Abstract In this Chapter, the author specifically focuses on the socioeconomic and arsenic related situation that prevailed in three villages of Meherpur in December, 2000. Meherpur has been noted as one of the arsenic-affected districts in Bangladesh. The three selected villages of Bagoan Union (an administrative unit) were Bagoan, Tarangar and Dholmari; No significant remedial action was found to be taking place there at the time of the survey, albeit the Department of Public Health Engineering (Meherpur) informed that these three villages were badly affected. The findings show that Taranagar was the worst-hit in terms of arsenic contamination in ground water among the three villages.

In order to address the arsenic mitigation issue, it is necessary to gather information on the socio-economic and arsenic-related situation of the areas concerned. At the initial stage of the study three villages, namely, Bagoan, Dholmari and Taranagar were selected. Bagoan was quite a big village, with an area of 1555 acres, and a population of 3685 (estimated from Population Census of Bangladesh 1991). Taranagar was a medium sized village, with an area of 488 acres and around 2023 people. Dholmari was rather a small village, with an area of 325 acres and a population of about 560. The author carried out sample surveys among the heads and 'female heads' of households of these villages in order to find out their socio-economic and arsenic related situations. The respondents were selected through simple random sampling. There were about 830 households in Bagoan, 440 households in Taranagar and 110 households in Dholmari. At least 20% of households from each village were included in the sample and therefore, there were 45 respondents from Dholmari, 95 from Taranagar, and 166 from Bagoan. The data were collected for this particular survey in December, 2000. SPSS software was used to analyze the data. The findings are discussed below.

3.1 Socio-economic Situation

All the villages under the present study were predominantly agricultural. All the respondents' families (except one from Bagoan) drank water from tube wells and used the same for cooking. The respondents have been living in these villages for 30 years or more. All the respondents were more or less aware of the arsenic

© Springer Japan KK 2017   27
W. Akmam, *Arsenic Mitigation in Rural Bangladesh*, New Frontiers in Regional
Science: Asian Perspectives 16, https://doi.org/10.1007/978-4-431-55154-6_3

Table 3.1 Distribution of the number of persons in the respondent households

Village	2–3 persons (%)	4–5 persons (%)	6–7 persons (%)	8 and above persons (%)	Average (persons)
Bagoan	31.30	43.40	25.30	00.00	4.4
Taranagar	17.00	63.90	17.00	02.10	4.61
Dholmari	46.70	40.00	13.30	00.00	4.06

Table 3.2 Monthly income of the respondent households

Village	0–2000 Taka (%)	2000–4000 Taka (%)	4000–6000 Taka (%)	6000– 10,000 Taka (%)	>10,000 Taka (%)	Average (Taka)
Bagoan	57.80	20.50	13.30	07.20	01.20	2982
Taranagar	55.30	17.00	8.60	12.70	06.40	4002
Dholmari	60.00	33.30	6.70	00.00	00.00	2400

problem. They all owned the houses they lived in, although the type of house may be different. Nearly all of the respondents enjoyed the facility of having the water source (tube wells) within their home compounds. It may be mentioned that several households may live within a single compound, and a single tube well may serve as their water source. In Bangladesh, it is a matter of prestige concern, if women have to go to other houses for water. Therefore, people, no matter how poor, tried to sink a tube well within their own compound.

The average number of members in household in Bagoan, Taranagar and Dholmari were 4.4, 4.61, and 4.06 respectively. Seventy four point seven percent of the households in Bagoan had ≤5 persons, and the same was true for 80.9% and 87.7% of the households in Taranagar and Dholmari respectively (see Table 3.1).

The income of the residents of the three villages varied between 1500 Taka and above 10,000 Taka per month. However, 57.8% of the people in Bagoan, 55.3% of those in Taranagar and 60% of those in Dholmari earned 2000 Taka or less. In Bagoan, 20.5% of the respondents belonged to the 2000–4000 Taka monthly income group. Seventeen percent and 33.3% of those in Tarranagar and Dholmari (respectively) belonged to the same income group. Never-the-less, the average income level was much higher in Taranagar (4002 Taka) than in Bagoan (2982 Taka), and Dholmari (2400 Taka) (see Table 3.2).

The inhabitants of the three villages under study were mainly agriculturists (those who cultivated their own land) and agricultural laborers (those who worked on land owned by others for daily wage). There were of course a few businessmen and service holders (e.g., teachers, family planning workers etc.). The highest number of agriculturists was found in Taranagar (87%), and the lowest in Bagoan (37.35%). In Dholmari, 73.33% of the respondents were agriculturists, and 13.33% were day laborers. More than 34% of the respondents in Bagoan were engaged in agriculture, along with other occupations and 22.89% of them were agricultural laborers (see Table 3.3).

As the villages of my study were mainly agricultural, land holding was significant in understanding socio-economic condition of the arsenic affected people under study. The people in Taranagar were the highest average landowning people

Table 3.3 Distribution of the respondents belonging to different occupational groups

Village	Agriculture (%)	Agriculture + others (%)	Business (%)	Day laborer (%)
Bagoan	37.35	34.94	04.82	22.89
Taranagar	87.00	06.50	01.06	04.90
Dholmari	73.33	13.33	00.00	13.33

Table 3.4 Distribution of respondents in different land possessing categories

Village	00 Bighas (%)	00–02 Bighas (%)	02–04 Bighas (%)	04–06 Bighas (%)	06–10 Bighas (%)	>10 Bighas (%)	Average (Bighas)
Bagoan	24.01	18.10	22.90	13.20	08.40	08.50	7.43
Taranagar	21.30	17.00	17.00	14.90	06.40	23.40	10.83
Dholmari	13.30	06.70	20.00	26.70	26.60	06.70	5.97

Table 3.5 Distribution of the households in different average schooling years categories

Village	0 Year (%)	0–2 Years (%)	2–5 Years (%)	5–8 Years (%)	8–10 Years (%)	>10 Years (%)	Average (years)
Bagoan	24.10	15.70	21.60	21.70	13.30	03.60	4.21
Taranagar	04.30	26.7	42.70	19.10	02.10	02.10	4.20
Dholmari	26.60	20.00	13.30	33.40	06.70	00.00	3.47

[10.83 Bighas (1 Bigha = 0.33 acre = 1227.8 m^2)]. Bagoan (7.43 Bighas) followed Taranagar, and Dholmari was the poorest (5.97 Bighas). Landlessness was highest however, in Bagoan (24.01%), and the lowest in Dholmari (13.3%) (see Table 3.4).

Education (schooling) is another important socio-economic variable, which determines peoples' capabilities and actions. In Table 3.5 we find the average schooling years (ASY) of respondent households, which gives us a snapshot of the literacy situation. 'Schooling years' refers to the number of years spent in schools, colleges, universities etc. for acquiring formal education (excluding the extra years spent by an individual for repetition in one grade). 'Average schooling years of household' was calculated by adding the schooling years of all the members of each household (aged 12 and above) and then dividing the sum total by the number of members (aged 12 and above).

Regarding ASY also, we find Dholmari (3.47 ASY) lagging behind Taranagar (4.20 ASY) and Bagoan (4.21 ASY). Although the average ASY in Bagoan and Taranagar appeared to be almost the same, there were more illiterate people in Bagoan than in Taranagar (see Table 3.5).

3.2 Arsenic Related Situation

Findings of the survey show that the percentages of tube wells that had been tested for arsenic in Bagoan, Taranagar and Dholmari were 26.66%, 65.96% and 20% respectively. Among those tested, the percentages of contaminated wells were 77.77% in Bagoan, 80.64% in Taranagar, and 66.66% in Dholmari. Twenty percent

Table 3.6 Distribution of the diseases the arsenic-related patients suffered from

Village	Melanosis (%)	Melanosis + Keratosis (%)	Melanosis + Keratosis + Gangrene (%)
Bagoan	46.87	53.13	00.00
Taranagar	72.72	18.18	09.10
Dholmari	33.33	33.33	33.33

of the respondents in Dholmari, 38.55% of the respondents in Bagoan and 70.21% of the respondents in Taranagar had arsenic related patients in their households.

Most of the identified arsenicosis patients in the three villages had Melanosis – 46.87% in Bagoan, 72.72% in Taranagar and 33.33% in Dholmari. In Bagoan, most patients were suffering from Melanosis and Keratosis (53.13%). Those who suffered from Gangrene were not likely to recover at all. It was not possible to identify those who were suffering from cancer as a result of drinking arsenic contaminated water. So, we cannot be certain about the actual number of patients. In the three villages of our study, patients below the age of 7 years were identified, which points to the possibility that the level of arsenic toxicity in the contaminated water in these villages was very high (see Table 3.6).

There are some studies (e.g., Sarkar 1999; Hassan and Das 2000; Chowdhury 2000), which point towards social problems caused by arsenicosis. The skin lesions resulting from melanosis and keratosis appear to be leprosy like symptoms. Leprosy is an infectious disease that damages skin and nerves. People are afraid of this disease and abandon those who show such kinds of symptoms.

Even when the patients are diagnosed as patients of arsenicosis (which is not contagious), they are isolated and socially maltreated. Women with the disease are sometimes divorced and sent back to their parental homes. Men with arsenicosis often lose their jobs. Unmarried women with family members suffering from arsenicosis (even if they themselves are not patients) are likely to remain unmarried, as nobody would offer to marry them. Poor women affected by the disease are not hired for domestic work. It has been observed that the people of an affected village are considered outcasts and people from non-affected villages do not want to establish marital relationship with them. According to Nahar (2000), women are more negatively affected than men. In the survey carried out in Bagoan, Taranagar and Dholmari, respondents who had arsenic related patients in their households, said that they did not face any social problem because of arsenicosis. However, they themselves suffered from inferiority complex because of their overt disease and tried to hide themselves as much as possible.

The respondents were asked regarding the actions taken by the government and other NGOs in order to mitigate the problem in their villages. According to the respondents, no action had been taken for Dholmari, either by the government or by NGOs and relatively more measures had been taken for Taranagar (see Table 3.7). The respondents could not mention the name of the NGO working in their villages, they only spoke of a Japanese woman, the name of whom they could not pronounce properly. However, all the respondents said that neither the government, nor NGOs did anything to actually supply safe water to them. Their activities were limited to

Table 3.7 Percentages of the respondents answering in a positive direction on actions taken by the government and/ or NGOs

Village	Government (%)	NGOs (%)
Bagoan	22.89	00.00
Taranagar	27.66	80.85
Dholmari	00.00	00.00

Table 3.8 Distribution of the alternative options chosen by the women respondents

Village	Deep tube well (%)	Filtering tube well water (%)	Dug well (%)
Bagoan	48.20	50.60	01.20
Taranagar	59.57	40.43	00.00
Dholmari	86.67	13.33	00.00

testing of tube wells and marking them with red (signifying not safe for drinking), and green (signifying that they were safe).

The respondents were in favor of deep tube wells and filtering arsenic contaminated water as alternative options (see Table 3.8). The main reason for choosing deep tube wells in Taranagar might be the influence of the NGO workers, who told the villagers that deep tube wells were safe. However, deep tube well aquifers are not available everywhere, and the cost of searching for aquifers is very high. The villagers chose filtering of tube well water because in that case they do not have to go outside their homes to get the water. But they were not aware of the environmental hazards involved in filtering. Only one respondent in Bagoan said that his family drank water from dug wells in fear of arsenic related diseases. This person was educated and had arsenic related patients in his household. The respondents were not at all willing to drink from ponds even after filtering, as they believed that pond water was polluted and could not be purified by filters because the ponds and river (Bhairab) in these villages became almost waterless during the dry season.

The respondents were asked about their willingness to pay to get safe drinking water. It has been observed earlier that 55–60% of the respondents of each village earned less than 2000 Taka (US$ 40). So, it was really difficult for them to pay some extra money to buy safe water. Table 3.9 shows the willingness of the respondents to pay per month for the purpose of getting safe drinking water. In the table, we find that the percentage of people not willing to pay at all is the highest in Bagoan. More than 86% people in Dholmari, and 70% in Taranagar were willing to pay, but the amount was as small as 50 Taka (US$1) per month.

In order to establish safe water systems, there is an initial cost. The respondents were asked about the amount of money they were willing to donate at the initial stage of constructing a safe water supply system. In Table 3.10 we find that 55–72% of the inhabitants were not willing to donate at all. Relatively, a high percentage of respondents were willing to donate at the initial stage in Taranagar (44.7%), the seemingly worst affected and the richest village among the three (in terms of arsenic related patients). It appears that the respondents considered the problem as if not of their own but of the government to make provisions for arsenic-safe water.

People who cannot spend money for establishing and maintaining safe water systems can give voluntary labor. Among the respondents, 91.57% in Bagoan, 100% in Taranagar and 86% were willing to give voluntary labor in order to get safe water.

Table 3.9 Distribution of the respondents' willingness to pay (per month) to get safe water

Village	00 Taka (%)	00–50 Taka (%)	50–100 Taka (%)	100–200 Taka (%)	>200 Taka (%)
Bagoan	44.60	38.60	08.40	07.20	01.20
Taranagar	29.8	66.00	02.10	02.20	00.00
Dholmari	13.30	66.70	20.00	00.00	00.00

Table 3.10 Distribution of the willingness of the respondents to donate at the initial stage

Village	00 Taka (%)	0–500 Taka (%)	500–1000 Taka (%)	1000–2000 Taka (%)	2000–5000 Taka (%)
Bagoan	72.3	12.00	04.80	03.60	07.20
Taranagar	55.30	31.90	12.80	00.00	00.00
Dholmari	66.70	13.3	13.3	06.70	00.00

Table 3.11 Distribution of the women's willingness to walk

Village	00 Min (%)	0–5 Min (%)	5–10 Min (%)	10–15 Min (%)	Average (Min)
Bagoan	25.30	63.90	03.60	07.20	03.80
Taranagar	12.80	74.50	12.80	00.00	02.55
Dholmari	00.00	93.30	06.70	00.00	02.58

In Bangladesh, it is mainly the womenfolk who cook and procure the water for cooking and drinking. Therefore, having to walk to bring safe water means extra load of work for them (Nahar 2000). The 'female' heads of households were asked regarding the distance they were willing to walk to procure safe water. Some female heads of household considered their going out of house for procuring water as a prestige-hampering matter. Therefore, they answered that they were not willing to walk at all to bring water. Table 3.11 portrays the response of the female heads household (or wife of male respondents) in the three villages, regarding their willingness to walk and the average (in minutes) willingness to walk. In the table, we find that all the female respondents in Dholmari and 87% of those in Taranagar were willing to walk. The percentage of women who were not willing to walk at all was the highest in Bagoan (25.30%). Most of the respondents were willing to walk for 5 min – 63% in Bagoan, 74.5% in Taranagar and 93% in Dholmari. The average willingness to walk in Bagoan, Taranagar and Dholmari were however, 3.8 min, 2.55 min and 2.58 min respectively.

The 'female' heads/wife of male heads of household were requested to give their preferences (on a hierarchy) regarding how they wanted to get their drinking/cooking water. All of them gave the following hierarchy:

1. Tap water at home
2. Filtered tube well water

3. Tap water at walking distance
4. Filtered river/pond/dug well water

In the hierarchy, it is obvious that they preferred options, which allowed them to get the water at home, without requiring them to go outside. While trying to find out a solution to the problem, we must keep in mind the preferences expressed by these respondents.

3.3   Association/Correlation Between Variables

Some cross tables were prepared in order to see how pairs of some of the key variables were correlated/associated. Table 3.12 shows whether these variables were associated (or correlated) or not. In order to test the relationships, Pearson's Correlation Coefficient was used in cases where both the variables were quantitative (marked as *), and in cases where one of the variables was qualitative (marked with ** in the table), Chi-square test was applied. In Table 3.12, significant association (SA) can be observed between land holding (Land) and arsenicosis patients (Pa) in Dholmari ($\chi^2 = 5.79$, $d.f.= 1$) and Bagoan ($\chi^2 = 15.75$, $d.f.= 1$). No significant association between these variables was observed in Taranagar. However, significant association (SA), was observed between income level (Inc) and arsenicosis patients (Pa) in Bagoan ($\chi^2 = 5.79$, $d.f.= 1$), but not in Taranagar and Dholmari. No significant association between average schooling years (ASY) and arsenic patients in household (Pa) was found in Taranagar and Dholmari. However, these variables were found to be significantly related in Bagoan ($\chi^2 = 67.5$, $d.f.= 2$). Table 3.12 also shows that land holding did not affect either willingness to pay every month (WP) to get safe water, or willingness to donate at the initial stage (Don) in any of the three villages under this study. In Bagoan and Taranagar, significant correlation (SC) was observed between Income (Inc) and willingness to pay (WP), but not in Dholmari. On the other hand, significant correlation between income and donation at the initial stage was observed in Bagoan and Dholmari, but not in Taranagar. Significant correlation (SC) at the 0.01 level was observed between education of the respondent (Edu) and willingness to pay (WP) in Bagoan, but not in the other two villages. No significant correlation ((NSC) between education (Edu) and donation at the initial stage (Don) was found in any of the villages. Although educational level of women (WE) and their willingness to walk to get safe drinking water were not significantly correlated in any of the villages under this study, income of household (Inc.) and willingness to walk (WW) were found to be significantly correlated at the .01 level in Bagoan. In the other two villages, these variables were not significantly correlated. Having patients in household seems to have significant impact on the willingness to pay and willingness to donate at the initial stage in Bagoan [($\chi^2 = 20.64$, $d.f.= 2$); ($\chi^2 = 17.48$, $d.f.= 1$)]. However, those variables were not significantly associated in Taranagar and Dholmari.

The findings derived from Table 3.12 show that the prevalence of arsenic patients was so widely distributed in Taranagar that the variable was not significantly affected

Table 3.12 Correlation/association between variables

Village	**Land/Pa	**Inc/Pa	**ASY/Pa	*Land/WP	*Land/Don	*Edu/WP	*Edu/Don	*Inc/WP	*Inc/Don	**Pa/WP	**Pa/Don	*WE/WW	*Inc/WW
Bagoan	SA at .01 level	SA at .01 level	SA at .01 level	SC at .01 level	NSC	SC at .01 level	NSC	SC at .01 level	SC at .01 level	SA at .01 level	SA at .01 level	NSC	SC at .01 level
Taranagar	NSA	NSA	NSA	NSC	NSC	NSC	NSC	SC at .05 level	NSC	NSA	NSA	NSC	NSC
Dholmari	SA at .05 level	NSA	NSA	NSC	NSC	NSC	SC at .01 level	NSC	SC at .05 level	NSA	NSA	NSC	NSC

*Pearson's Correlation Coefficient; **Chi-square Test Value

by factors like ownership of land, income and average schooling years. However, in Bagoan, prevalence of patients was significantly associated with all the three factors mentioned above meaning that higher level of land holding, income and ASY would reduce the possibility of having arsenic-related patients in household. In Dholmari, only land holding was associated with prevalence of arsenic-related patients in household. This implies that mitigation policies should concentrate more on supplying safe water to those households that belong to the lower levels of income, land holding and ASY.

Table 3.12 also shows that willingness to pay was significantly correlated with income in most cases, which was very much logical. Therefore, while formulating a policy for arsenic mitigation, these issues had to be brought under consideration.

Chapter 4
Available Safe Water Options

Abstract This chapter concentrates on the different safe water options that are available for the arsenic affected people – deep tube wells, dug (shallow) wells/ring wells, treatment of tube well water, treatment of surface water, and rainwater harvesting. Taking into account performances of the different options based on different criteria – e.g., cost, capacity in terms of households served, satisfaction of WHO/Bangladesh standard for arsenic, WHO/Bangladesh standard for bacteria, whether produces toxic sludge or not, whether safe for retention of ground-water, socially acceptable or not, whether easily operable or not, geographically feasible or not, the author analyzes pros and cons of each of the available options.

Information on different options of providing safe water to the arsenic-affected people was gathered through studying different research reports and publications collected using internet communications and through personal contacts. Scrutinizing the gathered information, it was found that there were broadly five types of safe water options, namely,

1. Deep tube-wells;
2. Dug (shallow) wells;
3. Treatment of tube well water;
4. Treatment of surface water; and
5. Rainwater harvesting

While trying to find an alternative system to provide safe water for the arsenic affected people, cost-effectiveness of the options, their acceptability, the level of convenience offered to the people concerned, and its environment-friendliness have to be considered. BAMWSP recommended four non-chemical based options for short-run purposes, e.g., (1) Pond Sand Filter, (2) Rainwater Harvesting, (3) Dug well and (4) Deep Tube well (conditionally approved) (BAMWSP 2002). The Technological Advisory Group (TAG) of BAMWSP also recommended Stevens Institute Technology Filter, Tetrahedron Filter, Sono three-pitcher Filter, BUET Activated Alumina Filter and Alcan Enhanced Activated Alumina Filter. However, filtering processes involved sludge production, which needed expert management to avert possibility of another environmental problem.

W. Akmam, *Arsenic Mitigation in Rural Bangladesh*, New Frontiers in Regional Science: Asian Perspectives 16, https://doi.org/10.1007/978-4-431-55154-6_4

Similar recommendations were made by the participants at the International Workshop on Arsenic Mitigation held in Dhaka, during January 14–16, 2002. As long-term options, the participants suggested the adoption of the short-term options that are proven to be safe and sustainable technologies, and/or introduction of piped water supply (see Ahmed 2002).

4.1 Deep Tube Wells

Some thinkers consider sinking deeper tube-wells is the best solution to solve the problem. The WHO is taking such an initiative. However, deep tube well aquifers are not found everywhere and the process of searching for such aquifers is very costly. There is no guarantee that the deeper aquifers are not going to be contaminated. Observations made in various areas of Bangladesh and in West Bengal, India show that deep tube wells were also becoming contaminated (Prothom Alo 2001). Therefore, constant monitoring is required. Moreover, it is very costly to install deep tube wells, and has the risk of environmental deterioration. Besides, if there is actually a relationship between excessive groundwater extraction and the release of arsenic in ground water, this step may make the situation even worse. For Taranagar, deep tube well could be an option, although the above statements make it more or less prohibitive.

4.2 Dug Wells

Dr. Dipankar Chakraborti and his colleagues (DCH and SOES, JU 2000) have found after testing that water of dug (shallow) wells are less contaminated than shallow tube well water. He argues that before the introduction of tube wells as sources of drinking water, people in this region were habituated to drink water from dug wells and there is no evidence of them suffering from arsenic related diseases. However, Professor Meer Husain (through e-mail communication) questions the validity of Dr. Chakraborti's sampling and argues that it is not safe to drink dug well water, (as dug well water is also ground water) before testing on the hand dug wells by experienced professionals. Despite this warning from Meer Husain, Dr. Ain-un-Nishat, a renowned water expert in Bangladesh, opines that for the time being, dug wells may be an alternative, if the wells are protected with rings and toilets are kept at a reasonable distance (Prothom Alo 2001; Ahmed 2002). Some probable explanations for lower concentrations of arsenic in dug wells as mentioned by Ahmed (2002) are the following:

1. Precipitation of dissolved iron and arsenic may be caused by the oxidation that occurs in dug well water due to the agitation that takes place during withdrawal of water and also as a result of dug wells' exposure to open air.

2. The top layer of ground water is accumulated in dug wells, which "by percolation through aerated zone of the soil" (p. 4) is replenished every year by rain and surface water that is safe from arsenic.
3. "The presence of air and aerated water in wells can oxidize the soils around dug wells and infiltration of water into wells through this oxidized soil can significantly reduce the concentration of arsenic in well water (p. 4)."

In a study by BRAC (Ahmed, H. S. 2000), it has been suggested that a combination of dug well and surface water filter could be a good option, lessening the threat of bacteria to a great extent. Together, this 'joint option' satisfies the Bangladesh standard for arsenic and both Bangladesh and WHO standard for bacteria. Dr. Arif Mohiuddin Sikder (Sikder and Khan 2002) has developed a "Sanitary Dug Well", through which it is possible to get safe water directly from the facility; however, it does not satisfy the bacteria standard either for Bangladesh or the WHO. For Meherpur, ring/dug wells could be a good option. During my visit to the Department of Public Health Engineering of Meherpur in 2000, the author was informed that considering it as a good option for the area, a project was undertaken to install 40 ring wells in the affected areas of the district. The number was not at all sufficient, but their choice of option was taken into account for the area as such. However, cautious observations were to be made before the dug well water could be used for drinking. The water of the dug wells must be monitored while being continuously withdrawn for a few days, and kept under complete sanitary protection.

4.3   Filtered Arsenic Contaminated Tube Well Water

Treating the arsenic contaminated water is another widely considered method. There are various ways to treat the arsenic contaminated water. All these processes have their relative advantages and disadvantages. Never-the-less, most of these involve the management of dumping of the arsenic contaminated residuals, which may create yet another environmental problem in the long run. The methods of filtration which were considered by different NGOs and in this study include, Three Pitcher Filter, SIDKO Plant, Tube well Sand Filter, and Activated Alumina Filter (Ahmed, H. S. 2000; Grameen Bank 2000).

4.4   Treated Surface Water

Treatment of surface water is considered a good option for Bangladesh. There are plenty of rivers and ponds scattered all over the country. Every village has at least one pond. This plentiful water can be treated to remove the harmful elements and provide safe water to the affected people. Ms. Mortoza (by e-mail communication) has suggested (after reviewing all the available procedures) that installation of ultra-violet dis-infection plants was the best solution for Bangladesh. These plants could be installed at a cost of less than US$ 1.6 billion for providing treated water for all citizens of the country for a period of 15 years. The maintenance cost would be about US$ 5 million per year. The maintenance procedure was very simple. However, the people have to come to the water plant site to carry the water to their homes. Pond Sand Filter and Surface Water Filter are two other cheaper means of treating surface water (Ahmed, H. S. 2000). However, surface water is not available in dry seasons in some areas, and the ponds and rivers become turbid and in some cases, waterless. This is especially true in the case of our study villages. Social acceptability of treated surface water as an alternative to tube well water is very low. People have internalized in their minds that surface water should be avoided for drinking, in order to be safe from water borne diseases (Ahmed, H. S. 2000; Grameen Bank 2000). In order for pond sand filter and surface water filter to be effective, it is required that the water is free of fertilizers and other chemical contamination. Fulfilling this condition is very difficult in Bangladesh.

4.5   Rainwater Harvesting

Bangladesh has plenty of rains, almost throughout the year, except for the winter dry season. Many people are speaking for rainwater harvesting as a solution for providing potable water to the arsenic-hit people in Bangladesh. "Catchment" and "Storage" are two important aspects of the rainwater harvesting option. C. I. (corrugated iron) sheet roofs are used to collect water, which is collected in a tank through a gutter system. UNICEF has designed a system of rainwater collection, which is able to serve a community for 6 months. According to Sanchia Chowdhury (1998), a rainwater- harvesting system can be easily constructed, and is reliable. The water provided by the system is of good quality. However, the amount of average yearly rainfall is less than 1500 mm in the study area (Website of Mujibnagar 2015), making this option not suitable.

4.6   Pipeline Water Supply

According to Ahmed (2002), the ultimate goal should be to develop a piped water supply system, as it can supply water to the consumers at a close proximity and thus can compete with the existing system of tube well water; it is safe from external contamination; it provides better quality control; and as the required amount of water can be easily collected. However, it is very costly and difficult to be constructed in the rural areas where the population is scattered. To supply water for a population of about 420 households, the construction of a piped water supply system costs nearly 2,000,000 Taka. For piped water supply, the source of water that

Table 4.1 Information on different safe water options

Name of option	Cost (in Taka)	Capacity in terms of households served	Satisfies WHO As. standard	Satisfies Bang. As. standard	Satisfies Bang. and WHO bacteria standard	Produces toxic sludge	Safe for ground water	Socially acceptable (taste)	Complex in operation	Geographically feasible
Deep tube well	5333	60	No	Yes	Yes	No	No	Yes	No	Yes
Dug well	2333	25	No	Yes	No	No	Yes	Yes	No	Yes
Dug well + surface water filter	8583	25	Yes	Yes	Yes	No	Yes	Yes	No	Yes
Sanitary dug well	5000	40	Yes	No	No	No	Yes	Yes	No	Yes
Rain water harvesting	6133	1	Yes	Yes	Yes	No	Yes	No	Yes	No
Pond sand filter	4666	60	Yes	Yes	Yes	No	Yes	No	Yes	No
Surface water filter	1000	1	Yes	Yes	Yes	No	Yes	No	No	No
UV water disinfection	100,832	800	Yes	Yes	Yes	No	Yes	No	Yes	No
Three pitcher filter	1000	1	No	Yes	Yes	Yes	Yes	Yes	No	Yes
SIDKO plant	12,666	75	Yes	Yes	Yes	Yes	Yes	Yes	Yes	Yes
Tube well sand filter	900	20	No	Yes	Yes	Yes	Yes	Yes	Yes	Yes
Activated alumina filter	41,000	30	Yes	Yes	Yes	Yes	Yes	Yes	No	Yes

Sources: Ahmed H.S. (2000), Grameen Bank (2000), Sikder and Khan (2002), Waterhealth Inc. (2001), Ahmed (2002)

has higher capacity, e.g., U.V. water disinfection plants, deep tube wells, would be more suitable provided that they are feasible for a certain area.

In Table 4.1, information on the different safe water options that have been considered in this research is provided. These have been collected from several sources, e.g., Ahmed H.S. (2000), Grameen Bank (2000), Sikder and Khan (2002), Waterhealth Inc. (2001), Ahmed (2002). Some data have been collected through personal (e-mail) communication with the researchers, e.g., Md. Jakariya, Dr. Arif Mohiuddin Sikder, Dr. Abul Hasnat Milton.

The cost of each option is roughly calculated by the following formula:

$$\frac{\text{Establishment cost} + \left(\text{Yearly maintenance cost} \times \text{Longevity}\right)}{\text{Longevity}}$$

For the community-based options, longevity has been considered as 15 years. It has been reported in the Ahmed, H. S. (2000) study and in Prothom Alo (2001) that the three-pitcher filter, which is capable of supplying safe water for only one household lasts for only 3 months. Therefore, for a single household, four facilities would be required to supply safe water for a year. Surface water filter, which also provides bacteria-safe water for one household can work for several years, but require new 'permeable candles' after each 2 years of time.

Chapter 5
A Model for Attaining Guidelines to Select Optimal Safe Water Option(s) and to Specify their Locations

Abstract In this chapter a multi-objective mixed-integer optimization model is built. The model constitutes a set of structural equations, which define constraints of the mixed-integer programming model. The policy mix, in order to supply safe water for the people living in the affected area, is heavily dependent on the optimal solution of the model. The optimization considers not only an optimum selection among the set of safe options but also the optimality of assignment of the selected safe option all over the village. The optimality is mostly dependent on multi-objects, namely the number of people who can get safe water on the specified standard, the total cost necessary for construction and maintenance, environmental friendliness of the selected safe options, total hours consumed by the people living in the village to get the safe water, geographical feasibility, ease in operation etc. In developing the model a *Primitive Paretian Optimization* approach was adopted, i.e., an approach in which one of the multi-objects is to be maximized (or minimized) keeping all the other multi-objects at various specific levels jointly with several structural equations of demand, supply, etc.

5.1 Skeleton of the Model

The skeleton of the model is shown in Flow Chart 5.1.

Following the skeleton of the model, at first a set of structural equations that defined the constraints of the mixed-integer programming model was formulated. The policy-mix that was recommended to supply safe water for the people living in Taranagar village was determined by the optimal solution of the model. The optimization considered not only an optimum selection among the set of safe options, but also the optimality of assignment of the selected safe option across the village. The optimality was dependent on multiple objectives, for instance, the number of people who were able to get safe water on the specified standard, the total cost necessary for construction and maintenance, environmental friendliness of the selected safe options, total hours consumed by the people living in the village to get the safe water, geographical feasibility, and ease in operation. As mentioned earlier, we adopted in this article a primitive Pareto optimization approach, namely, an approach in which one of the objectives was to be maximized (or minimized) while keeping all the other objectives at various specified levels, jointly with several structural

© Springer Japan KK 2017

W. Akmam, *Arsenic Mitigation in Rural Bangladesh*, New Frontiers in Regional Science: Asian Perspectives 16, https://doi.org/10.1007/978-4-431-55154-6_5

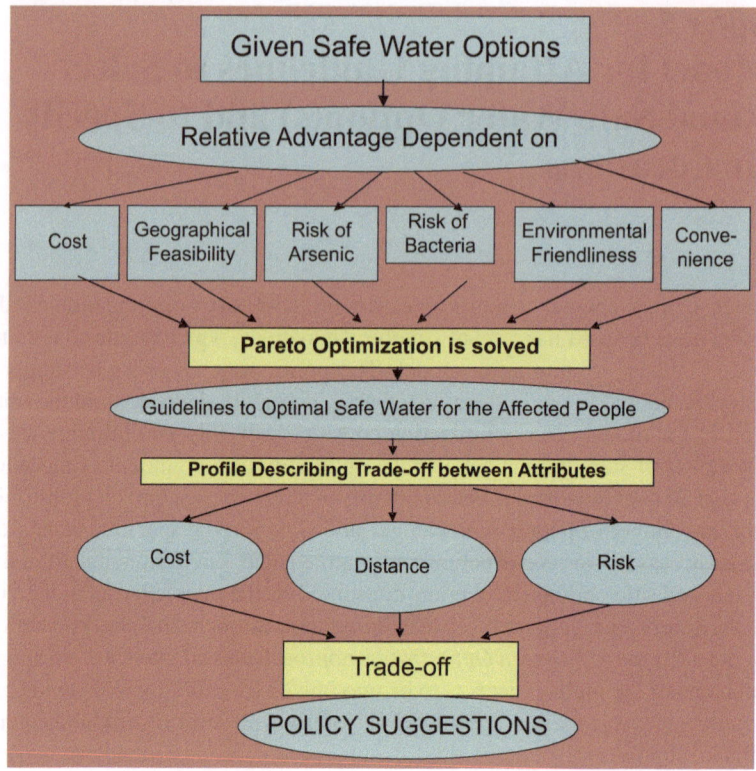

Flow Chart 5.1 Skeleton of the model

equations of demand and supply, etc. An outline of the model is shown in Flow Chart 5.1.

5.2 Mathematical Equations

5.2.1 Demand for Water

The first equation requires that the number of people who are living in zone $i(i \in I)$, and who can get safe water from the safe water option $k(k \in K)$ available for them in zone $j(j \in I)$ must be equal to or greater than the total number of people living in zone i:

$$\overline{X}_i \leq \sum_{k \in K} \sum_{j \in I} \sigma_{ijk} X_{ijk}, \ \left(i \in I \right), \tag{5.1}$$

where I is the set of zone indices $\{1, 2, 3,..., N\}$, and N the number of zones, K the set of safe water option indices $\{1, 2, 3,..., T\}$, and T the number of safe water options. For example, $k = 1$ means safe water is provided by deep tube well, $k = 2$ means dug well, $k = 3$ means a combination of dug well water and surface water filter, $k = 4$ means sanitary dug well, $k = 5$ refers to rainwater harvesting, etc. The parameter $\sigma_{ijk} = 1$ if it is acceptable for the people living in zone i to walk to zone j to get water by safe water option of type k; if not $\sigma_{ijk} = 0$ $(i, j \in I, k \in K)$. As most of the women have said that they were willing to walk for only 5 min (approximately 360 m), the maximum distance to the water was fixed at 360 m. X_{ijk} is the number of people who were living in zone i and got safe water through safe water option k in zone j $(i, j \in I, k \in K)$, and $\overline{X}_i$ is the number of people living in zone i.

5.2.2   Capacity of Safe Water Option

The second equation requires that the total number of people who get safe water through safe water option k in zone j to be equal to (or less than) the capacity of the safe water option k located in zone j:

$$\alpha^k Y_j^k \geq \sum_{i \in I} \sigma_{ijk} X_{ijk}, \quad j \in I, k \in K, \tag{5.2}$$

in which, α^k is a capacity coefficient of safe water option of type k in terms of the maximum number of households that can be provided with safe water through $(k \in K)$, and
  Y_j^k the number of safe water options which are to be built, or which already exist in zone $j (j = I)$.

5.2.3   The Number of Households that Can be Provided with Safe Water

The third equation defines the (least) number of people in the village who can get water at a certain maximum permissible risk level q in terms of a certain risk category c:

$$X^{cq} = \sum_{k \in K} \sum_{i \in I} \sum_{j \in I} \sigma_{ijk} S_{qk}^c X_{ijk}, \quad q \in Q^c, c \in R, \tag{5.3}$$

in which $S_{qk}^c = 1$ if safe water option of type k can provide safe water at the risk level of q in terms of risk category c, or at a risk level lower than that $(k \in K, q \in Q_c, c \in R)$. The assessment of when $S_{qk}^c = 1$ is not only based on information regarding safe water technology, but also on the geological information about arsenic risk exposure in the zone where the water supply is located. $Q^c = \{1, 2, 3, \ldots M^c\}$ is a set of indices of the risk level in terms of risk category c, and M^c is the number of risk levels. For example, when $c = 1$ and $q \in Q^1$, $q = 1$ means water consumption is totally risk free in terms of the arsenic problem, $q = 2$ means risk free in terms of arsenic contamination on the basis of the WHO standard (0.01 mg/l), $q = 3$ means risk free in terms of the Bangladesh Government guideline (0.05 mg/l), $q = 4$ refers to the risk level at which the people who drink the contaminated water for 10 years or more are likely to be attacked by arsenicosis with a probability of 0.8, etc. When $q \in Q^2$, $q = 1$ means risk free from bacteria on the basis of WHO guideline (<1 colony/100 ml), and $q = 2$ refers to risk freeness from bacteria on the basis of the Bangladesh standard (<10 colonies/100 ml), etc. X^{cq} is the number of people in the village who can get safe water at the risk level of q in terms of category c, or a lower risk level than that. R is the set of risk categories with $c \in R$. For example, $c = 1$ refers to the risk of arsenic in the water, $c = 2$ refers to the risk of bacteria in the water, $c = 3$ refers to the risk of toxic sludge being generated, $c = 4$ refers to the likelihood of the lowering of the ground water level, $c = 5$ refers to the likelihood that the water is not acceptable (in terms of the taste), $c = 6$ refers to the likelihood of too much complexity in operation, making the option less convenient and $c = 7$ refers to the risk of the particular option not being geographically feasible.

5.2.4　Necessary Cost

The annual cost of each option is roughly calculated as the ratio of the sum of the establishment cost, and the accumulated maintenance cost over the lifetime, divided by the expected life of the option. For the community based options, on the basis of the rough information gathered from different sources, the expected life has been assumed to be 15 years. It has been reported by Ahmed H S (2000) that the three-pitcher filter, which is capable of supplying safe water for only one household, lasts for only 3 months. Therefore, four facilities would be required to supply safe water for a year to a single household. A surface water filter, which also provides bacteria-safe water for one household, can work for several years but require new 'permeable candles' after every 2 years of operation.

The fourth equation defines the total cost necessary for the establishment of safe water options in the village:

$$C = \sum_{k \in K} \sum_{j \in I} c_k Y_j^k, \tag{5.4}$$

in which C is the total cost necessary for the establishment of safe water options, and C_k refers to the maintenance and depreciation cost of the safe option of type $k(k \in K)$.

5.2.5 Accessibility

The fifth equation defines the sum of the distances people must walk to get water:

$$D = \sum_{k \in K} \sum_{i \in I} \sum_{j \in I} d_{ij} \sigma_{ijk} X_{ijk},$$ (5.5)

in which D refers to the total distance people must walk to get water in the village, and dij refers to the walking distance between zones i and j.

5.2.6 Paretian Optimization

Having described the problem and the technological, geographical and economic characteristics of the people, the policy guidelines found through the multi-objective programming model, using the Pareto optimization theory are discussed below.

5.2.7 Multiple Objectives

There were three main objectives: (1) to lower the cost necessary for the establishment of safe water options irrespective of whoever pays for it, (2) to increase the convenience of the safe water options proposed for the people in the village in terms of accessibility, and (3) to increase the number of people who can be provided with safe water at the maximum permissible arsenic level, or a lower level than that.

Generally speaking, if the preference of the village as a whole could be identified over the trade-off among the above three objectives, we would only solve the following optimization problem:

$$\max_{\{X_{ijk}, Y_i^k\}} F\left(C, D, X^{11}, X^{12}, \ldots, X^{1Q^1}, \ldots, X^{GQ^1}\right)$$ (5.6)

subject to Eqs. (5.1), (5.2) and $F(C, D, X^{11}, X^{12}, \ldots, X^{1Q^1}, \ldots, X^{GQ^G}$ in which $F(\ )$ refers to the preference of the village as a whole.

However, it is assumed that it is very difficult to identify such a preference function since the people living in the village have never been involved in such public decision-making, and have difficulty in understanding what the process should be for agreement on how to solve the arsenic problem. Some of the villagers may not even understand the problem itself. So, the author suggests that it would be better if a profile of several alternative varieties of safe water options is proposed to the people. But beforehand, the people must understand the problem itself, as well as what options are open to them, so that it would be most probable that they could

Fig. 5.1 Trade-off
between Walking Distance/
Risk and the Minimized
Cost of Safe Water Options
$k (k \in K)$.

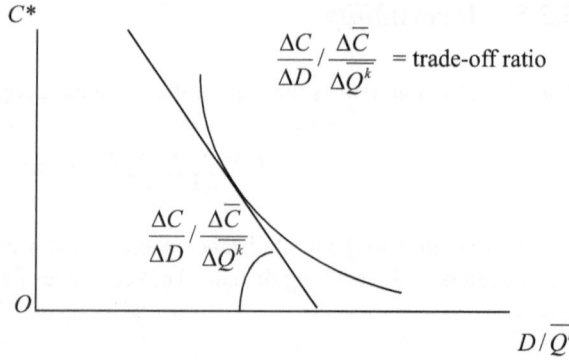

arrive at an agreement on the choice of certain safe water option(s) through a trade-off between risk, cost and distance, meeting the standard of the so-called informed consent (see Fig. 5.1). The model presented here provides a guideline of optimal options from which they would be able to choose the option that they feel is the most convenient and suitable to them. The people are to be provided with the information regarding the risks involved, the costs required for specific options and the level of convenience they will enjoy. They would also be offered a guideline of optimal options, making combinations of different levels of risk alleviation, cost and convenience. In order to reach a decision, of course, a collective decision mechanism based on informed consent must be established.

5.2.8 Cost Consciousness

If the people involved are anxious about how much money they are required to pay, the safe water options should be an optimal solution to the following problem:

Problem 1

$$\min_{\{x_{ijk}, y_j^k\}} C,$$ (5.7)

subject to Eqs. (5.1) and (5.2),

$$X^{\overline{cQ^c}} \; \gamma^c \cdot \overline{P} \; (c \in R),$$ (5.8)

and

$$D \le \overline{D} \cdot \overline{P},$$ (5.9)

in which, $\overline{Q^c}$ is the maximum permissible risk level in terms of category $c \in R$ $\left(1 \le Q^c \le \overline{Q^c}\right)$, $X^{c\overline{Q^c}}$: total number of people who can get water at the maximum permissible risk level in terms of category c, γ^c the percentage of the households that can get safe water, meeting a certain risk level in terms of category $c(0 \le \gamma^c \le 1)$, $\overline{P}$ the total number of households in the village; and $\overline{D}$ the maximum acceptable average walking distance.

In the beginning, it would be useful to set $\overline{D}$ equal to a sufficiently large value. By doing so, it would be possible to simulate in order to find out the cost required to solve the problem at a certain risk free level, irrespective of the average walking distance.

Figure 5.1 shows how the government of Bangladesh and the inhabitants of Taranagar must make a trade-off between walking distance D and cost of safe water options (C). In order to reduce the walking distance, the people must pay more for the option, and to reduce the cost, the people must walk longer distances. The same figure also shows the trade-off between risk and cost of safe water options. With stricter risk levels $\left(\overline{Q}\right)$, the cost of the safe water option (C) is going to be high, and if the strictness of the risk levels are relaxed, the cost of the option can be reduced. A decision has to be made collectively regarding the level of risk that would be acceptable to the people, and the cost that the people and the government can bear.

The minimized cost, C^*, must be dependent on a set of parameters which specify the maximum acceptable average walking distance, percentage of risk free people with certain risk level in terms of all risk categories whose values that are fixed at certain levels in advance, namely,

$$C^* = \psi\left(\gamma^1, \ldots, \gamma^G, \overline{Q^1}, \ldots, \overline{Q^G}, \overline{D}\right), \tag{5.10}$$

Equation (5.10) explains how the people and the government must make a trade off among the factors of permissible maximum risk level of category c, $Q^c\left(c \in R\right)$, the percentage of the people who are free from the arsenic problem at that risk level, $\gamma^c(c \in R)$, the maximum acceptable average walking distance, $\overline{D}$, and the minimized cost, C^*. If the parameters, γ^c and $\overline{D}$ could be fixed, e.g., politically, we can draw the ($G + 1$)- dimensional profile, which shows how the people must make a trade-off between the (minimized) cost and the maximum permissible risk level. If $\gamma^c(c \in R)$, are all set equal to 1, the ($G + 1$)- dimensional trade-off shows how the people must compromise regarding the cost and maximum permissible risk levels that are applicable for all the people living in the village. This compromise will make the inhabitants realize how much money they must pay if they have to be risk free in all the risk categories at certain levels. If the parameters $\gamma^c(c \in R)$, and $Q^c\left(c \in R\right)$ are given, e.g., being based on a certain exogenous standard, for example, following the WHO criterion, Eq. (5.10) can show how the people in the village must make a trade-off between the accessibility and the cost to be borne by the people.

5.2.9 Accessibility Consciousness: When the Cost would be Subsidized

Assuming that a subsidy is given from outside the village, an alternate problem can be formulated as follows:

Problem 2

$$\min_{\left\{X_{ijk}, Y_j^k\right\}} D,$$ (5.11)

subject to Eqs. (5.1), (5.2), (5.(8)), and

$$C \leq \bar{C},$$ (5.12)

in which, $\bar{C}$ is the amount of subsidy given from outside the village.

The minimized average walking distance D^* can be shown as follows:

$$D^* = \phi\left(\bar{C}, \overline{Q^1}, \dots, \overline{Q^G}, \gamma^1, \dots, \gamma^G\right).$$ (5.13)

Using Eq. (5.13) we can draw a trade-off between accessibility and the maximum permissible level of risk in terms of a certain category when other parameters are fixed.

5.2.10 Risk Exposure and the Minimum Cost

Problem 2 may show that there is no feasible solution if $\bar{C}$ is too small to meet a certain lower maximum permissible level of risk. The extreme case is given by the case that $Q^c = 1$ for all $c \in R$. Such a case implies that some people have to be exposed to risk in terms of a certain category with some risk, other things being equal. In other words, people must compromise with less than 1 for some c ( ). The number of persons who must be exposed to risk of a certain category ( ) at the specific risk level ( ) can be calculated by solving the following problem:

Problem 3

$$\max_{\left\{X_{ijk}, Y_j^k\right\}} X^{c'q'},$$ (5.14)

subject to Eqs. (5.1), (5.2), (5.9), (5.12); and

$$X^c \overline{Q^c} \geq \gamma^c \cdot \bar{P},$$ (5.15)

for a certain $\overline{Q}^c$ $(1 \leq \overline{Q}^c \leq \overline{Q}^c$ $)$ for all c $(c \in R, c \neq c')$,in which it may be useful to set $\bar{D}$ equal to the sufficiently large value in Eq. 5.9. So, $\bar{P} - \left(X^{c'q'} \right) *$ is the number of persons who would be exposed to the risk of category c' at the risk level of q' in which $\left(X^{c'q'} \right)$ is an optimal solution to Problem 3.

The model specified above could be used as a part of a social device through which the community may arrive at an agreement on the solution for the arsenic problem. In the next chapter, the simulation results derived from the model are discussed.

Chapter 6
Simulation Results

Abstract The basic aim of this chapter is to describe the results of simulations carried out to test the efficiency of the model developed in Chap. 5. The main objective of carrying out the simulations was to find the efficient loci of the trade-offs between several competing objectives: the cost that the affected people have to pay to get safe water, the risk they will be exposed to and the distance they would have to cover to bring water from the facility site. Taking into account various other constraints also, such as environmental friendliness, geographical feasibility, complexity in operation, and social acceptability, the most efficient locus of the trade-off was found in the establishment of 18 dug wells in specific zones, and providing each household with a surface water filter with which they can filter the dug well water before drinking. The inhabitants of Taranagar would have to walk about 2 min in order to bring the dug well water from the closest facility to their own houses. This option will satisfy all constraints except the WHO standard for arsenic (Case 1). If we take into account all the constraints (Case 2), the optimal option would be the establishment of 15 community-based Activated Alumina Filters in specific zones. However, the total cost will increase more than fourfold. The trade-off point for cost and distance for Case 1 is at an average distance of 58 m, and for Case 2, the trade-off point is at an average distance of 67 m.

In the simulation results, the researcher expected to find guidelines for a compromise between risk, cost and distance. Due to the indivisibility of the safe water options, the variables in Chap. 5 had to be integers. So, the original model is a mixed-integer programming model. However, it had too many integer variables to be solved directly with the mathematical programming software package that was used (LINGO). Consequently, a non-linear cost function was introduced, which roughly approximated the original step-wise cost function in that an original 0 or 1 integer variable was replaced by a continuous variable between zero and one or zero and two,..., and the functional form still had the relative advantage for the continuous variable to take 0 or 1 or 2... in terms of the cost. The original cost function was approximated by a hyperbolic function, for which non-linearity was maximized by making the power as small as possible in order to yield a relative advantage (see Fig. 6.1). Using this approximation the author was successful in getting feasible solutions for several cases: cost minimization without any constraints (referred to as

© Springer Japan KK 2017
W. Akmam, *Arsenic Mitigation in Rural Bangladesh*, New Frontiers in Regional
Science: Asian Perspectives 16, https://doi.org/10.1007/978-4-431-55154-6_6

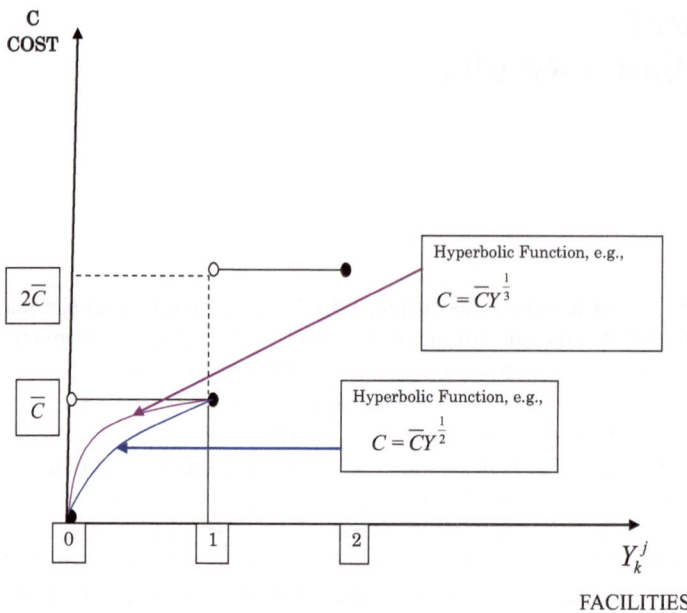

Fig. 6.1 Approximation by hyperbolic function

the basic case), cost minimization with all the constraints mentioned earlier and cost minimization with all constraints except the WHO standard for arsenic.

6.1   Simulation Cases Indifferent to Distance

The results showed that with an objective to minimize cost in the basic case, the optimal option was a tube well sand filter (TSF). In this case, the average cost per household was about 45 Taka per year, with an average distance (which the affected people will have to walk to get safe water) of 141 m, which was less than 2 min walking distance. However, the selected option did not satisfy the WHO standard for arsenic, the toxic sludge constraint or the ease in operation constraint. With all constraints except the WHO arsenic standard taken into account (Case 1), satisfaction of the arsenic standard in Bangladesh, the Bangladesh and WHO standard for Coliform bacteria, geographical feasibility, social acceptability in terms of taste of the water and complexity involved in operation of the systems, the optimal safe water option was a dug well combined with a surface water filter. For this case, the average cost per household was 354 Taka, with an average distance of 146 m (about 2 min walking distance).

Table 6.1 Simulation cases indifferent to distance

| Types of cases | Cost per year (Taka) | | Distance (m) | | Options selected | Number of facilities required |
	Average (per household)	Total	Average	Total		
Basic case	45.3	19,800	141.29	61,745	Tube well Sand filter (TSF)	22
Satisfying all risk constraints except WHO arsenic standard (Case 1)	353.53	154,494	145.9	63,798	Dug well + surface water filter	18
Satisfying all constraints (Case 2)	1407.32	615,000	168.18	73,495	Activated alumina filter	15

6.2 Simulation Cases for Trade-Off between Cost and Distance

In the social survey carried out among the villagers, the respondents overwhelmingly opted for facilities that would allow them to get the water at home. They wanted the distance reduced as much as possible. However, most of them were not in a position to fund the establishment of community based safe water options. Therefore, the profile of the trade-off between cost and distance for the last two cases mentioned in Table 6.1 needs to be shown. Simulations were carried out, gradually decreasing the total distance found in Case 1 and 2 (see Figs. 6.2 and 6.3, respectively) in order to see how the distance must be compensated by the cost. After carrying out several simulations with different distances, we found the trade-off curves for the case satisfying all constraints, and the case satisfying all but the constraint of the WHO arsenic standard. In Fig. 6.2 and Table 6.2, we see that for the latter case, the trade-off point was around the total distance of 25,519 m, with an average distance of approximately 58 m, and the cost increased significantly from that point, as the distance decreased. For the case satisfying all the constraints (Case 2), a significant increase in the cost occurred at the point of 40% of the original total distance, which was 29,398 m, or about 67 m on average per household (see Table 6.3 and Fig. 6.3). In this case, reduction of the distance up to 50% without increasing the cost was possible just by redistribution of the location of facilities due to indivisibility of the facilities; the capacity of the facility was not fully used at up to 50% of the locations. Therefore, in those cases, the distance that people were required to walk to fetch safe water was quite small compared to the initial cost of the facility as well as the cost necessary to satisfy all (or almost all) the constraints. These simulation results show that by simply relocating the facilities in different zones, it was possible to reduce the distance to some extent. The results also depict

Table 6.2 Cost and distance trade-off (Case 1)

Total distance (m)	Average distance (m)	Percentage of distance (original: 65,628 m) (%)	Cost per year (Taka)	Number of facilities required
63,798	145.90	100	154,494	18
57,418.2	131.39	90	154,494	18
51,038.4	116.79	80	154,494	18
44,658.6	102.19	70	154,494	18
38,278.8	87.59	60	154,494	18
31,899	72.99	50	154,494	18
25,519.2	58.40	40	171,660	20
19,139.4	43.80	30	188,826	22
12,759.6	29.20	20	240,324	28
6379.8	14.60	10	326,154	38

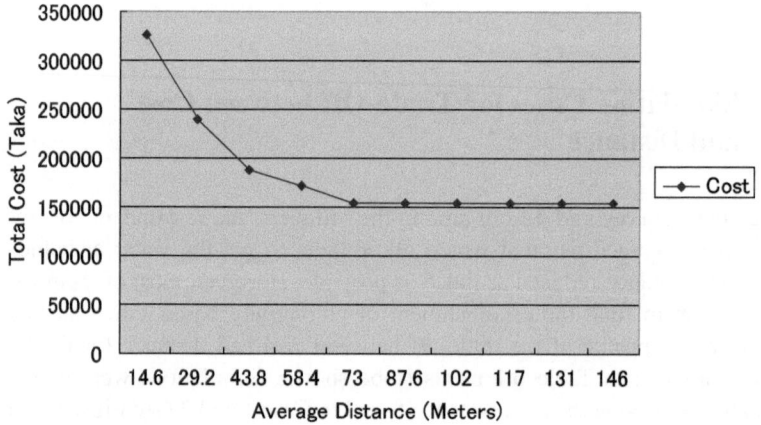

Fig. 6.2 Trade-off between cost and distance (Case 1)

Table 6.3 Cost and distance trade-off (Case 2)

Total distance (m)	Average distance (m)	Percentage of distance (original: 73,495 m) (%)	Total cost per year (Taka)	Number of facilities required
73,495	168.18	100	615,000	15
66,145.5	151.36	90	615,000	15
58,796	134.54	80	615,000	15
51,446.5	117.73	70	615,000	15
44,097	100.09	60	615,000	15
36,747.5	84.09	50	615,000	15
29,398	67.27	40	656,000	16
22,048.5	50.45	30	697,000	17
14,699	33.64	20	902,000	22
7349.5	16.82	10	1,476,000	36

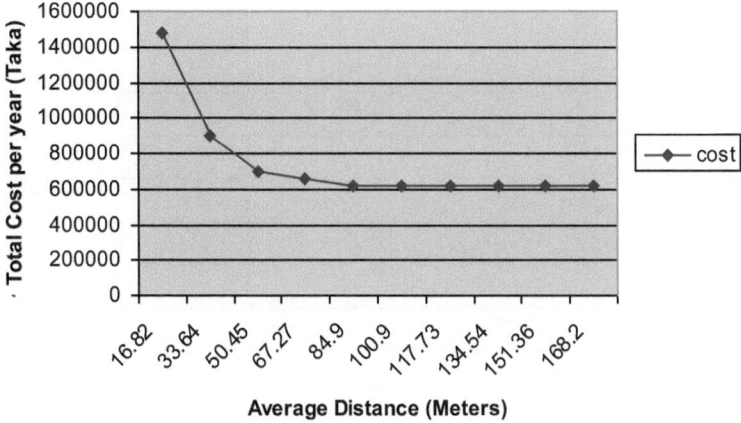

Fig. 6.3 Trade-off between cost and distance (Case 2)

the increased cost of reducing the distance, so that the affected people could decide for themselves whether they would pay for the additional decrease in the distance or not.

In the case satisfying all risk constraints, including WHO standard for arsenic (Case 2), the optimal option chosen was Activated Alumina Filter. In this case, the cost increased considerably (to 1407 Taka), with an average distance of 168 m, which was a little more than 2 min walking distance. This is shown in Table 6.3. The difference in the distance in Case 1 and Case 2 was the result of the difference in capacity of the options. The capacity of a dug well with a surface water filter was 25 households, whereas the capacity of an Activated Alumina Filter was 30 households. Therefore, the number of facilities required in case of the former option (dug well plus surface water filter) was 18, whereas only 15 facilities with an Activated Alumina Filter were required to be constructed in order to serve the total village population. If 18 facilities were distributed all over the village, the average walking distance would naturally be less than in the situation where there were only15 facilities.

6.3   Simulation Cases for Trade-Off between Cost and Risk Level

Figure 6.4 has been drawn using data from the Basic Case, Case 1 and Case 2, which shows that the cost is very low (only about 45 Taka per household per year) for the basic case which only satisfied the Bangladesh maximum arsenic standard, and not the WHO standard for arsenic, nor the permissible level of bacteria. If we choose to implement the bacteria standard along with the Bangladesh arsenic standard, the cost per household per year will be approximately 354 Taka. However, the cost will drastically increase if we choose to adopt the arsenic standard of WHO, resulting in a cost of more than 1407 Taka per household per year.

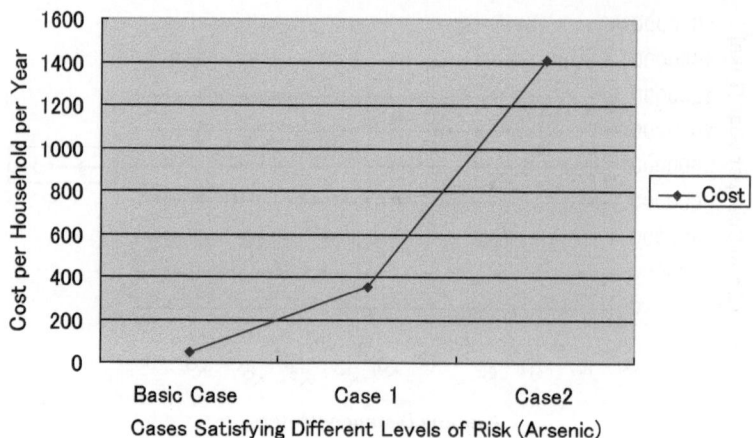

Fig. 6.4 Trade-off between risk and cost

Chapter 7
Mechanisms to Convince People to Adopt Safe Water Options

Abstract This chapter concentrates on ways to convince the affected people about the importance of drinking water from safe water sources and the mechanisms by which it might be possible to make them change their old habit regarding drinking water, and retain the new habit. Bringing about change in the habit of an individual is a very difficult task. According to the scholars in social psychology, the change in habit occurs in various steps.

7.1 Steps in Change of Habit

Step 1. Awareness of the new situation: People must be aware of the situation in which they are to change their habit. For our specific purpose, people must be aware of how they are exposed to the risk of arsenic poisoning, and the severity of the situation.

Step 2. Interest in that situation: The people must have interest in the affair – they must realize that it is them, who are going to be affected by this poisoning, and that it is in their own interest that they must know what is going on, and how they can be saved from the risk of arsenicosis.

Step 3. Evaluation of what is being persuaded: This stage is very crucial, as the people evaluate all the options they have regarding changing their habit, which they are being advised to do. The people will evaluate the cost (monetary, physical and psychological) and efforts they will have to make for switching to this new habit. The whole process has to be evaluated by the people as worthwhile, if they are to make any change in habit.

Step 4. Trial: After evaluating positively, the people will try the new methods they are being persuaded, at a very small scale. Here, the mental evaluation is tested against the actual experience (Read 1975).

Step 5. Adoption: After being convinced by the trial, the people are likely to adopt a certain safe water option.

Step 6. Maintenance of the new habit: Upon being satisfied with the results of adopting the new method, the people are likely to maintain that habit.

© Springer Japan KK 2017

W. Akmam, *Arsenic Mitigation in Rural Bangladesh*, New Frontiers in Regional Science: Asian Perspectives 16, https://doi.org/10.1007/978-4-431-55154-6_7

7.2   Communication Processes

The communication processes that are involved for our purpose consist of:

(a) Informing
(b) Instructing
(c) Persuading

While informing has a short-range learning goal, instructing involves a long-range learning goal, which is supposed to stimulate the receiver toward some additional activities. In case of persuasion, the receiver is expected to "yield to the point of view advocated by the persuader" (Schramm 1971, p. 44).

The first two processes involve clearing of four hurdles (Schramm 1971):

Attention: The message to be conveyed must attract the attention of those who are targeted;

Acceptance: After gaining attention, the message has to be accepted. Acceptance depends on the face value of the message and on the credibility of the person/source who is conveying the message;

Interpretation: The targeted person, after accepting a message will interpret it on the basis of his/her stored up experience and built-in values; and

Storing: The interpreted message will thereafter be stored for future use.

In the process of persuasion, the receiver is required to yield the message or point of view advocated by the persuader and act in the direction he/she is persuaded. In this process, credibility of the persuader is very crucial. Often a threat over which the receiver has no defense (but not so much as to arouse panic), guilt appeal, or even ridicule can help in yielding desired change in behavior. In Bangladesh, a strategic extension campaign on rat control used this tactic (by distributing posters that show trapped men in cages and rats laughing at them from outside), and was highly effective (Adhikarya 1989).

Cross-pressure on a specific target person can be built by convincing him/her that the two or more persons/groups he/she trusts opine in the opposite direction to his/her opinion, therefore, he/she should also change his/her opinion. This is called "strain toward consistency" (Schramm 1971). This tactic is also worth trying.

In Bangladesh, a social marketing approach has been very successfully applied in the field of family planning (Atkin and Meischke 1989) and in the use of oral rehydration saline for diarrhoea patients. Currently, a similar type of approach is being applied for bringing about the change in people's habit of drinking and cooking with arsenic contaminated tube well water. Social marketing involves designing, implementing and controlling of programs that seek to proliferate acceptability of a certain social idea, and or practice among targeted groups (Kotler 1975). More simply put, it is a strategy to change behavior, utilizing advances in marketing skills and communication technology through action framework and integrated planning (Kotler and Roberto 1989).

7.3 National Communication Strategy

At the initial stage of this crisis, a national communication strategy based on the social marketing technique was adopted to disseminate necessary information to the masses in the country on how they can address the problem (UNICEF 1999). In the first phase of the mitigation efforts (in a limited number of areas), the strategies used to build awareness among the people included radio-T.V. commercials, dramas, meetings with community leaders, workshops for health workers and other service providers, school-teachers and religious leaders, meetings with the villagers, distribution of leaflets, posters etc. (BAMWSP 2002). The local governments and NGOs were involved in different stages of the program. All the tube wells in the affected areas were tested and the people were informed whether it was safe for them to drink water from the tube wells, which they used. The mitigation authorities helped them in choosing a safe water option. School meetings, village meetings and meetings with locally elected leaders (bodies) were held to disseminate the messages of the arsenic problem to the people. Workshops were arranged to train the stakeholders. However, the people of our study area were only exposed to the messages sent through radio, T.V., and newspapers.

7.4 Awareness Level of the People

It is natural to assume that the affected people in Taranagar had already heard about the problem, as the percentage of households having arsenicosis patients was very high. However, in order to specify measures to be taken for these villagers in particular, it was necessary to know about the level of their awareness. It was expected that the 'aware' affected people would know that –

1. arsenic was a tasteless, colorless poison;
2. arsenic contaminated water was dangerous to health, and can produce many diseases, including gangrene and cancer which could lead to a relatively slow but miserable death;
3. high amounts of arsenic flowed from a mother's womb to her baby, and that growing children were very susceptible to arsenic poisoning;
4. both drinking and cooking water had to be safe, as the toxicity of arsenic did not lessen even after water was boiled.
5. there were specific safe water options that were suitable and available for them; they had to know about these specific options;
6. arsenicosis was not contagious and if already diagnosed, immediate treatment would be safe water, nutritious food and vitamins;
7. arsenic problem was not a curse of God, and could be overcome through adopting appropriate measures.

The awareness level of the respondents in Taranagar was measured on a five-point scale. To achieve the highest level, five, one had to (a) actually drink from safe water source; (b) demonstrate his/her full knowledge of the items mentioned above and (c) have willingness to donate for the establishment of safe water source, to pay every month for maintenance, and to give labor for the establishment and maintenance of the safe water source. While determining their level of awareness, financial and physical conditions of the respondents were taken into account. The awareness level of the respondents is shown in Table 7.1.

Table 7.1 shows that none of the respondents could reach the level of 5. The percentage of respondents achieving the level of 4 was also quite low (11.6%, n = 11). The percentage of those belonging to level 1 was quite high (26.3%, n = 25). Chi-square tests show that awareness level was not significantly associated with factors such as average schooling years ($\chi^2 = 0.051$; d.f. 1; significance level: 8.22), income ($\chi^2 = 2.44$; d.f. 1; significance level: 0.118), occupation ($\chi^2 = 0.873$; d.f. 1; significance level: 0.35) or patient in household ($\chi^2 = 0.073$; d.f. 1; significance level: 0.787). This finding was important in formulating ways of convincing people to drink from safe water sources.

In order to raise awareness of the people primary focus was on informing them about the actual high risk in continuing to drink contaminated tube well water. The next ordeal was to convince them about the necessity to change their habits regarding drinking and cooking water, to instruct them on how they could choose the option that would be optimal for them, to persuade them to switch to a safe water option, and to ensure that the new habit retained.

Making arrangements in which the establishment cost of the safe water options are paid by either the government, or some other donor agencies, and the maintenance cost are paid by the local people themselves (to have a sense of responsibility and a feeling of ownership) were proposed to convince the affected people. The risks involved in not switching to a safe water option, and the high costs, and environmental and technical problems related to establishing deep tube wells may be explained to them so that they could make a trade-off between cost, risk, distance and taste. Their doubts need to be subdued by clearly demonstrating each of the safe water options, especially those found optimal through simulation. Maintenance of coordination among all those involved in the mitigation project must be ensured

Table 7.1 Awareness level of the respondents

Awareness level	Frequency	Percentage
1.00	25	26.3
2.00	30	31.6
3.00	29	30.5
4.00	11	11.6
5.00	0	00.0
Total	95	100.0

through forming ward/union/village mitigation committees. A system discussed in Chap. 8 would help to maintain coordination between decision-making and implementation phases of the mitigation programs. Clearly spelled out responsibilities and specific authorities (committees) to which stakeholders are accountable may ensure proper functioning of any system.

7.5 Ways of Convincing People to Accept Alternative Safe Water Options

Villagers' proper understanding of the gravity of situation is a key towards bringing about change in their habit. In order to achieve the goal of convincing people to take the trouble of bringing water from a safe tube well or from an alternative safe water source the following steps need to be followed:

(a) Convey the messages in a convincing way stating how risky it is to continue to drink arsenic contaminated water, and providing them with information on the safe water options;
(b) Persuade them to accept a new option of drinking water in place of the old one; and
(c) Encourage them to retain the new habit.

Inauen, J., & Mosler, H.J. (2016) have suggested that behavioral change can be brought about more easily through a group of committed people who would (1) inform the people about the whole situation, (2) arrange mechanisms to remind them about the information through various memory aids, e.g. posters, pictures, etc., (3) help them in planning and executing the new work schedules in order to collect safe water for drinking and/or cooking and (4) obtain public commitment from the target group regarding their change in behavior. The village arsenic mitigation committee can play the role of 'the group of committed people' as discussed in the next chapter.

By following the steps mentioned above, keeping in mind the guidelines of behavior change discussed earlier, and by undertaking the following measures the goal of ensuring that all the villagers drink and cook with water from a safe water source can be attained:

1. *Raising Risk Awareness*: It is possible to raise risk awareness of the people through making them understand the high risks involved in continuing to drink arsenic-contaminated water without raising panic. They have to be told, for example, that internal cancers do not always show overt symptoms, and are rarely curable diseases and that they have the risk of having such cancer if they do not change their habit immediately.
2. *Applying Guilt Appeal*: Applying this mechanism [e.g., showing how the current young generation (the children) is likely to be diseased when they grow up because the older generation did not change their water drinking habit]. Ridicule

(showing how arsenic can be "victorious" over humans as a result of peoples' failure to change their habits) may also be successful in convincing the people that it is really necessary for them to switch to a safe water system. The change in habit made by those who are influential in the community, and those whom the people consider *knowledgeable and trustable* is likely to have a very positive effect. These *knowledgeable and trustable* persons may include teachers, religious leaders, and highly educated persons of the village who may be working as high-ranking officials outside the village.

3. *Efficient Cost Management*: For community based options, which require a substantial amount of money for construction, the establishment cost may be provided from outside the village, and the villagers have to pay for the maintenance.

4. *Continuous Monitoring*: A laboratory has to be established to test the arsenic in water samples (one laboratory sufficient for one whole district) in order to ensure regular monitoring. High school/college students can be engaged in collecting water samples from drinking water sources and monitoring them not only for arsenic but also for Coliform bacteria, and other substances that are detrimental to human health.

5. *Holding Village Meetings*: In order to explain the guidelines derived through simulation and the decisions made by the board of decision-makers regarding safe water options (see Chap. 8) to the villagers, village meetings or small group meetings can be arranged. These meetings are to be attended by the ordinary villagers, in the presence of Union Parishad (a local administrative authority unit) Chairman/members, teachers, religious leaders, highly educated people of the village (who are known as knowledgeable and are trusted by the villagers), wealthy people of the village, who may be interested in donating a handsome amount for the maintenance of safe water sources, and of course, field workers and students who will be working in the village for the mitigation purposes.

6. *Demonstration of Functioning of the Safe Water Options*: How each of the safe water options work must be clearly demonstrated. It might be a good idea to show video documentaries and arrange photo exhibitions. Provisions must be made so that all the questions in the minds of the people can be answered satisfactorily.

7. *Formation of Village Committees*: Village committees can be formed to coordinate the continuous process of mitigating the arsenic problem at local levels, so that everyone know what job they have to do and be responsible to the committee.

8. *Continued Motivational Communications*: Meanwhile, motivational communications must be continued through the media and through the trusted and influential personalities of the village.

In a nutshell, the whole process is shown in Flow Chart 7.1.

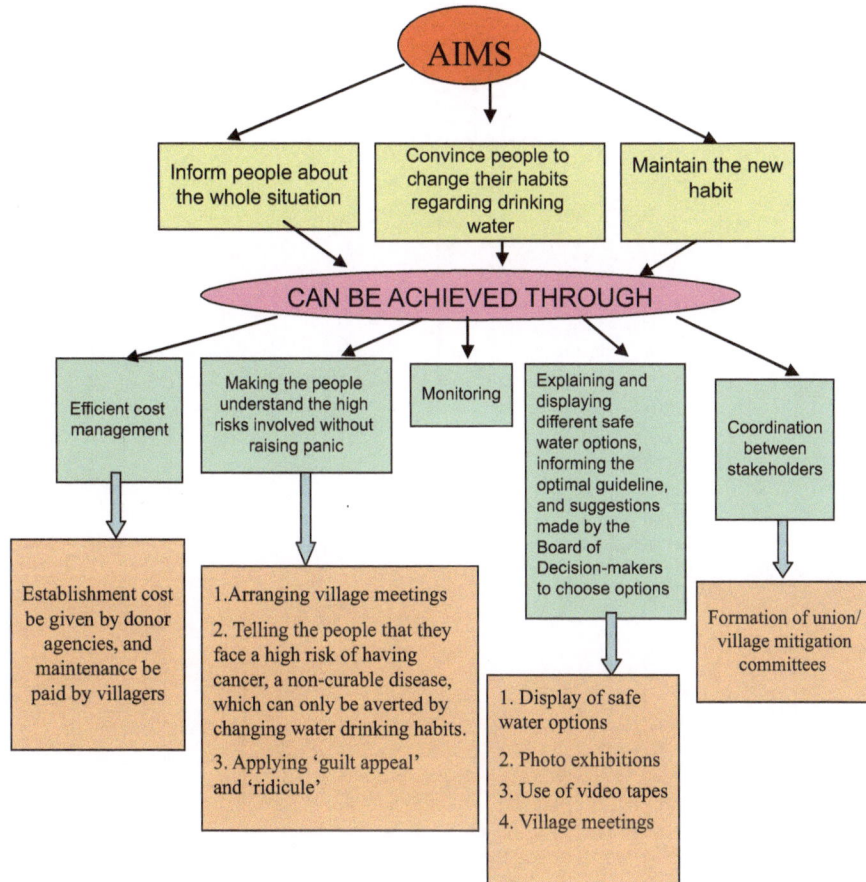

Flow Chart 7.1   Mechanisms for ensuring peoples' acceptance

Chapter 8
A System for Providing Safe Water

Abstract The aim of this chapter is to develop a system through which it would be possible to provide safe water to the people of the study area (Taranagar) through the most appropriate and optimal option available. At first a general system of seven stages is developed, followed by a relatively more detailed decision-making and implementation sub-systems.

A system may be "defined as a set of components interacting with each other and a boundary which possesses the property of filtering both the kind and rate of flow of inputs and outputs to and from the system" (Berrien 1968, pp. 14–15). By completing its voyage through a system, an 'input' transforms itself into a significantly and identifiably different 'output'. In a system, all parts of the structure are assumed to be linked either directly or indirectly and be "mutually responsive to one another" (Bates and Harvey 1975, pp. 29–30). The system developed in this chapter is necessarily a 'social system', which is defined by Talcott parsons (Parsons 1991, p. 24) as "a mode of organization of action elements relative to the persistence or ordered processes of change of interactive patterns of a plurality of individual actors."

The Government of Bangladesh is following a four step arsenic mitigation plan involving the Department of Public Health Engineering (DPHE), Directorate General of Health Services (DGHS) and Non-Government Organization NGO) partners (website of the DPHE, Bangladesh, 2016):

1. Awareness building (carried out by DPHE, DGHS and NGO partners);
2. Testing of tube well water and marking them with red and green to indicate their arsenic-safety (performed by DPHE and NGO partners);
3. Patient identification and management (performed by DGHS);
4. Providing alternative safe water options (carried out by DPHE and NGO partners).

In this study, a system is being proposed for actually providing safe water to the arsenic affected people. In order to decide upon the type of safe water option and to establish and maintain safe water options in different regions, a system has been developed, involving government offices, e.g., DPHE, DGHS etc., specialists, NGOs and the affected people themselves. It is to be noted here that the system being proposed is built in supposition that the National Screening Project (for this

© Springer Japan KK 2017
W. Akmam, *Arsenic Mitigation in Rural Bangladesh*, New Frontiers in Regional
Science: Asian Perspectives 16, https://doi.org/10.1007/978-4-431-55154-6_8

area) has been already completed. It is also assumed here that funds for the establishment of the safe water sources will be borne by the government or donor agencies via DPHE, and the villagers themselves will bear the maintenance cost.
Moreover, the government has to pay for the overall management procedures. The
villagers may be provided with subsidies and micro-credit to bear the establishment
costs, if sufficient funds for the establishment of facilities are not available.

8.1 System in General

The proposed general system is as follows:

1. First, scientists would try to find more cost-effective and feasible safe water
   options for the affected people.
2. These options will then be approved (or disapproved) by the Technical Advisory
   Group (TAG), formed by the DPHE.
3. A special committee will be formed (Technical Feasibility Committee) for each
   district to provide a guideline for region-specific feasibility of the TAG approved
   options. The committee will include specialists, regional administrators at different levels, local Department of Public Health Engineering (DPHE) officials etc.
   The committee will suggest the most feasible options for specific villages in the
   region.
4. Arsenic mitigation committees will be formed at union/ward level and village
   level. These committees will include representatives from specific villages, arsenic specialists, NGOs etc. Specific village committees will develop working
   plans to be approved by the Union Committees and the integrated working plans
   of unions/wards will be approved by the Regional Project Management Unit
   (RPMU).
5. The village committees will arrange village or small group meetings to explain
   to the villagers, the cost, efficiency, operation procedures etc. of the feasible
   options recommended by the region-specific committees. The villagers will have
   to come to a decision regarding selecting the safe water option, which they think
   would be most suitable for them. This decision has to be made before working
   plans are developed by the mitigation committees at different levels.
6. Having the work plans approved by the higher levels of the arsenic mitigation
   committees, and the RPMU, construction of the safe water sources will take
   place, with the help of village committees and support organizations.
7. Once the safe water sources are built and people start using them, regular monitoring of the sources has to be maintained by the village committees. In a nutshell, the system in general is shown in Flow Chart 8.1.

Flow Chart 8.1 System
in general

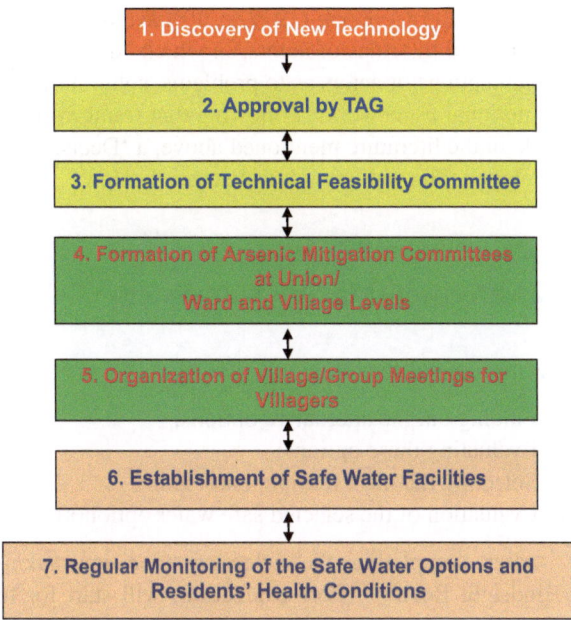

The general system discussed above can be further elaborated by dividing it into
two sub-systems, namely –

1. Decision-making Sub-system;
2. Implementation Sub-system.

In the process of decision-making and implementation of the establishment of
safe water system and their proper use, participation of those who will actually be
the beneficiaries is a prerequisite. Ample literature is available on how to involve the
citizens/the public in planning, and how to integrate the government and the community systems, etc. [e.g., please see Fagence (1977), Friend and Jessop (1977),
Iacofano (1990), Rondinelli (1977), Sewell and Coppock (1977)].

Iacofano (1990) has recommended some political and social process tools for
public involvement program design and management. Political processes involve
the specification of the purpose and the reason for public participation, level of participation required, type of commitments existing among public officials, persons
who should be involved, process of organizing participation etc. Social processes
include organizing the program, selecting the leaders and participants, arranging
settings, devising tasks, displaying information, providing resources and obtaining
feed back. As conventional means of citizen's participation in planning, Fagence
(1977) mentions the processes of exhibitions, public meetings/hearings, information documentation, questionnaire surveys, documentary reporting – the media,
ideas competitions, referenda, public inquiries etc.

Sewell and Coppock (1977, p. 9) have diagrammatized the process of policy-making and the roles of actors, in which "decision-making process is conditioned by perceptions of actors as to problems, solutions and responsibilities as well as by institutional framework." Following that framework, taking into account the points made in the literature mentioned above, a 'Decision-making Sub-system' has been developed. This sub-system is discussed below.

8.2 Decision-Making Sub-system

It is made up of the following phases:

1. Finding out the alternative options;
2. Evaluating these options;
3. Selecting the most feasible ones; and
4. Evaluation of the selected safe water option(s).

After completing the selection (decision-making) procedure, a phase of 'Hindsight Review' (Post–evaluation) will start for future reference. The 'Actor' who will evaluate whether the alternative options are feasible or not is the Technical Feasibility Committee described in the general system. The decision of this committee (i.e., the options, which are feasible for specific villages) will be forwarded to a board of decision-makers that consists of the representatives of village committees and government organizations [e.g., local Department of Public Health Engineering (DPHE)], support organizations who have previous experience in carrying out this type of task, and economists (who will help to decide on the optimal option). The analysis made by the board (e.g., trade-off between the competing objectives), which reflects upon on the decisions it makes, will be explained to the villagers, who should finally select the option which they would consider the best for themselves. Through the hindsight review the lessons learned in completing the project will be fed-back to the board of decision-makers, and if necessary, to the technical feasibility committee (see Flow Chart 8.2).

8.3 Implementation Sub-system

Once the options(s) are decided, work for implementation will begin. The central BAMWSP[1] will primarily distribute the funds for the implementation of projects. The Regional Project Management Unit (RPMU) of BAMWSP (or the DPHE) will

---

[1] BAMWSP (Bangladesh Arsenic Mitigation and Water Supply Project) was the national program set up in 1997 to mitigate the arsenic problem Bangladesh. However, the program ended in 2007 (Hanchett, and Monju, T.H. 2009). The 'system' developed in this study was structured in 2002 when almost all arsenic mitigation activities functioned via BAMWSP. As BAMWSP is not func-

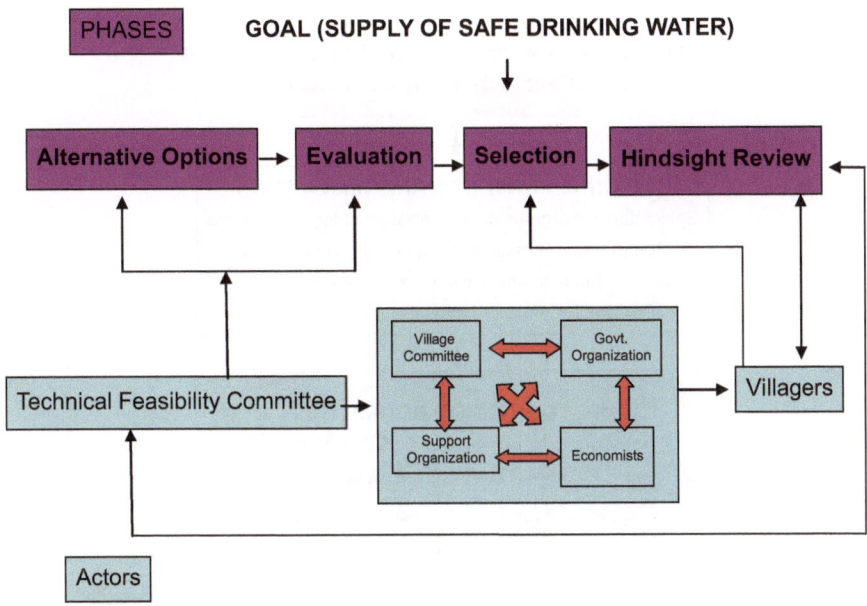

Flow Chart 8.2 Decision-making Sub-system

disburse funds for specific villages/unions (wards). The other functions of the RPMU include selection of the support organization having the experience of achieving success in this field, formation of implementation committee (support organization + village or union/ward committee), and organization of necessary training for the members of this committee. In case the villages of a union (an administrative unit made up of a cluster of adjoining villages; in municipal areas unions are known as 'wards') are not highly affected, and need only a few safe water facilities, union/ward committees will be able to handle the implementation work. The tasks of the implementation committee are mainly to organize village meetings in order to specify the exact location of facilities, construct the facilities, monitor the water quality, ensure that people drink from the safe water facilities and record data on people's health conditions regularly (see Flow Chart 8.3).

tioning at present, the 'system' may work through the government body of Department of Public Health Engineering (DPHE).

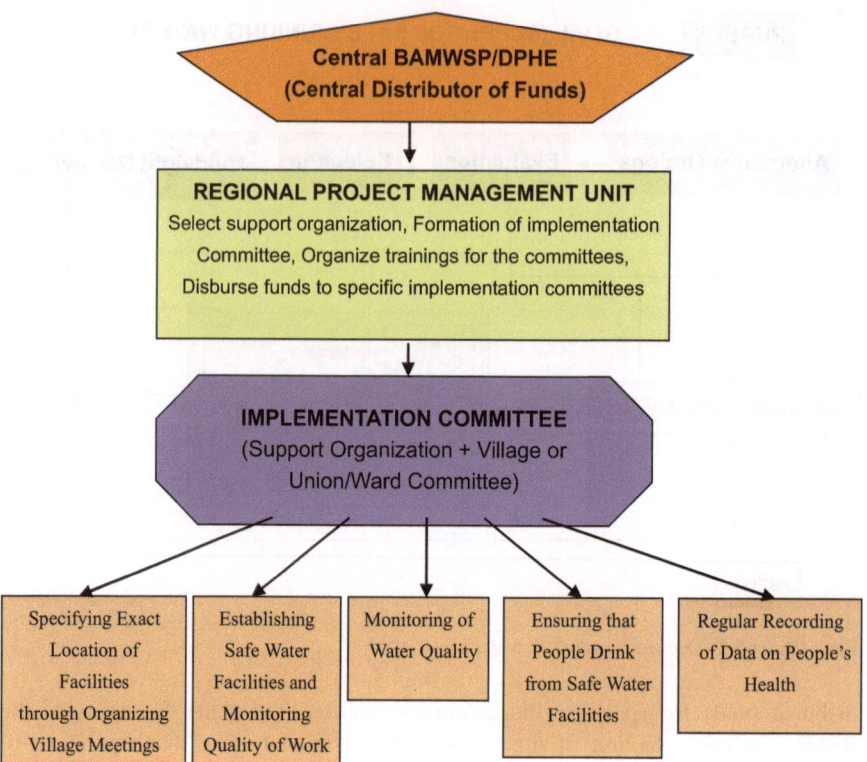

Flow Chart 8.3 Implementation Sub-system

Chapter 9
Taranagar from 2000 to 2012: Changes in Arsenic-Related Situation

Abstract Two follow-up surveys have been carried out in Taranagar to know about the arsenic-related developments in the village. Following the survey in 2000, a social survey was carried out in 2005 to observe the situation in this village after 5 years. Again another follow-up survey in the village was undertaken in 2011–2012. In this chapter changes in the arsenic-related scenario from 2000 to 2011 have been depicted. It has been observed that in 2005 arsenic-related situation had worsened in Taranagar as compared to 2000. Five people had died of arsenicosis. This situation raised awareness level of the people to some extent – yet, only half of the residents were drinking water from alternative water sources, such as dug wells and ring wells. In 2012 the percentage of households with arsenicosis patients reduced, meaning that awareness of arsenic had increased significantly and some interventions in combating the arsenic-related situation had taken place. However, only about half of the respondents interviewed in 2012 were sure that they were drinking from safe sources.

9.1 Collection of Data

This chapter focuses on the socio-economic and arsenic-related situation that was observed in Taranagar at three points in time – 2000, 2005 and 2011–2012. After 2000 two other follow-up surveys had been carried out in Taranagar to keep track of the arsenic-related developments that occurred in the village. The findings of these surveys have been presented in a way that allows comparison of the situations at three different periods. The same questionnaire with some minor changes was used to collect data at the three points in time. At least 20% of the estimated number of heads of household in Taranagar and their wives were interviewed (95 in 2000 and 175 in 2005 and 251 in 2012) with help of appointed interviewers. In rural Bangladesh, women mainly perform the task of collecting water for the household. For this reason, women's opinions on acceptable alternative safe water options were sought.

© Springer Japan KK 2017   73
W. Akmam, *Arsenic Mitigation in Rural Bangladesh*, New Frontiers in Regional
Science: Asian Perspectives 16, https://doi.org/10.1007/978-4-431-55154-6_9

9.2   Socio-economic Condition of Taranagar in 2000, 2005 and in 2011–2012

In 2005, significant change in the infrastructural/occupational features as compared to 2000 was not apparent in Taranagar. There were no concrete roads in the village, although there was a high school and a mosque. By 2012 nearly all the roads were concrete. However, the inhabitants still depended almost entirely on agriculture, with no recognized cottage industry. Taranagar did not have any river flowing adjacent to it. The closest river Bhairob was situated 2–3 km away and the water in the river was also not usable for drinking as the water was used for fermentation of jute. However, there was a big *beel* (depression) within the village, which contained water all year round in varying amounts. The area of the depression was almost 33 acres. This water body was entirely used for fishery, and for washing livestock; thus not suitable for the purpose of drinking/cooking. There were only seven ponds in the village, three of which became waterless in the dry season. Other ponds contained at least some water during the whole year. During the October-May period farmers of Taranagar used 109 shallow tube wells for irrigation purposes, with varied depths of 46–58 m (150–190 f). Almost all spewed water dangerously contaminated by arsenic. There were no deep tube wells installed within the village. All efforts made by the government and non-government organizations to sink deep tube wells proved to be futile. The major crops produced in the area were paddy, wheat, vegetables, potato, mustard, lentils, jute, sugar cane, onion, garlic, banana etc. Before going on to the comparison between the situations observed in the three periods of time regarding arsenic, the author would like to present a comparative scenario of three socio-economic variables: income, education and occupation.

9.2.1   Income

Table 9.1 portrays the income levels of the Taranagar residents in 2000, 2005 and 2012. It is observed that in 2005 the number of households with higher level of income reduced, while the percentage of the lowest income group remained almost the same. The average income per household had reduced significantly (from Taka 4498.94 to 3557.14). The reason could be that the productive power of the inhabitants had reduced as a result of arsenicosis. However, in 2012 the average monthly income had increased up to Taka 5055.58 and the households earning up to Taka 3000.00 or less had reduced to 47.8% from 64.6%.

Table 9.1 Monthly income of respondents in Taranagar in 2000, 2005 and 2012

Monthly income (in Taka)	2000		2005		2012	
	No. of respondents	Percentage	No. of respondents	Percentage	No. of respondents	Percentage
0–3000	61	64.2	113	64.6	120	47.8
3001–6000	15	15.8	46	26.3	86	34.3
6001–10,000	10	10.5	14	8.0	28	11.1
>10,000	9	9.5	2	1.1	17	6.8
Total	95	100.0	175	100.0	251	100.0

Table 9.2 Average schooling years of respondents' households

Average schooling years	2000		2005		2012	
	No. of respondents	Percentage	No. of respondents	Percentage	No. of respondents	Percentage
0–3	43	45.3	113	64.6	110	43.8
3.01–6	38	40.0	46	26.3	93	37.1
6.01–10	10	10.5	14	8.0	46	18.3
>10	4	4.2	2	1.1	2	0.8
Total	95	100.0	175	100.0	251	100.0

9.2.2   Education

Average schooling years of all the members of the respondents' household (those aged seven and above) were calculated in order to have an overall idea of educational situation in the village. The findings show that the mean average schooling years of respondents' households had increased a little (from 4.18 in 2000 to 4.49 in 2005). Table 9.2 portrays that a significant development in households' average schooling years had occurred during the study period (2000–2012). In 2000, 14.7% of the households had an average education of 6.01 years or more, where as in 2005, the percentage had reduced to 9.1% but in 2012 the percentage had increased up to 19.1. This reflects that more children are continuing school nowadays, which helped in building awareness about many necessary things such as arsenic contamination and the environment.

9.2.3   Occupation

Table 9.3 portrays the occupational situation of the respondent households in 2000, 2005 and 2012. It is observed that irrespective of the year of data collection, most households were found to be principally or partially dependent on agriculture. It is to be mentioned here that when farmers work on their own land and/or hire others' labor for cultivating their own land, they are said to be engaged in *agriculture*. The

Table 9.3 Occupation of the respondents' households

Occupation	2000		2005		2012	
	Frequency no. of respondents	Percentage	No. of respondents	Percentage	No. of respondents	Percentage
Agriculture	62	65.26	37	21.1	79	31.5
Agriculture and business	6	6.32	29	16.6	37	14.7
Agriculture, business and labour	0	00	2	1.1	1	0.4
Agriculture, business and services	0	00	1	0.6	0	00
Agriculture and labour	4	4.21	59	33.7	42	16.7
Agriculture and services	6	6.32	11	6.3	11	4.4
Business	5	5.26	12	6.9	21	8.4
Labor	0	00	20	11.4	52	20.7
Services	10	10.53	2	1.1	1	0.4
Services and labour	2	2.10	2	1.1	0	0
Business and labour	0	00	0	00	7	2.8
Total	95	100.00	175	100.0	251	100.0

laborers were mostly engaged in agricultural jobs. It is also observed that in 2005 most households had become dependent on more than one occupation. In 2012, however, the proportion of those dependent on *agriculture and labor* reduced significantly from 33.7 to 16.7% and the proportion of those dependent solely on labor increased from 11.4 to 20.7%. This points towards the fact that marginalization of ownership of land in Taranagar was progressing steadily.

9.3   Arsenic-Related Condition in Taranagar in 2000, 2005 and 2012

In 2000, all of the respondent households were using tube well water for drinking and cooking. Fifty six out of 95 of the respondents (58.95%) in 2000 had their tube wells tested, and among those tested 80.64% were found contaminated. More than 70% of the households had arsenicosis patients. Findings of the 2005 survey show (see Table 9.4) that more than 50% of the households were using dug well/ring well water for drinking, although there was complaint that water of ring wells often had bad odor. Moreover, they could not use dug well/ring well water for cooking, as the

Table 9.4 Kind of water used for drinking in 2005 and 2012

	2005		2012	
Water used for drinking	No. of respondents	Percent	No. of respondents	Percentage
Dug well water	50	28.6	39	15.5
Filtered tube well water	1	0.6	37	14.7
Ring well water	42	24.0	61	24.3
Tube well water	82	46.8	106	42.2
Pond water	0	0	5	2.0
Others	0	0	3	1.2
Total	175	100.0	251	100

safe water source was far away from their residence. Still 82 (46.8%) households used tube well water and six (7.32%) households used water from tube wells, which were yet to be tested for arsenic. Among those tested, 92.69% (n = 76) of the tube wells were found to be contaminated. Thus in spite of having patients in their households, a significant number of households continued to drink from contaminated tube wells. The reasons they mentioned were that they were habituated to drinking tube well water and did not want to change it; and that there was no alternative source of drinking water available to them. In 2005 more than 90% of the households had patients. Five patients had already died during the 5 year period. In 2012 the percentage of households with arsenicosis patients came down to 55.4%, meaning that awareness of arsenic had increased significantly and some interventions in combating the arsenic-related situation had taken place. However, only 49% of the respondents interviewed in 2012 were sure that they were drinking from safe sources and 6.4% did not know whether their sources were safe or not. More than 42% of the respondents continued to drink water from tube wells, 90% of which were contaminated. This situation raised some questions regarding the awareness level of the Taranagar inhabitants regarding arsenic-poisoning. The safe water sources they mainly used were dug wells, ring wells and arsenic filters. As Taranagar was situated in an area where average yearly rainfall is less than 1500 mm, rainwater harvesting was not a feasible option. The author of this book with the help of CWAJ established one dug well and two ring wells in Taranagar after the 2011–2012 survey had been completed. The author was informed by key informants residing in the village that during the years 2012–2017 some more safe water sources (SIDKO plants) had been built in the village.

9.3.1   Awareness of the Respondents

The DPHE (Department of Public Health Engineering) and NGOs (Kranti, EADS and Podokkhep) were involved in awareness raising programs in Taranagar. Under their projects all the tube wells in the area were supposed to be tested and the people were to be informed whether it was safe for them to drink water from the tube wells

Table 9.5 Willingness to pay in 2000, 2005 and 2012

Willingness to pay (Taka per month)	2000		2005		2012	
	No. of respondents	Percentage	No. of respondents	Percentage	No. of respondents	Percentage
00	36	29.3	1	0.6	68	27.1
0–50	55	44.7	153	87.4	172	68.5
50–100	1	0.8	16	9.1	10	4.0
100–200	3	2.4	3	1.7	0	0.0
200–300	0	00.0	2	1.1	1	0.4
Total	95	100.0	175	100.0	251	100.0

which they used. However, according to the respondents at least 12% of the tube wells used by respondents in 2012 had not been tested, and 12.7% of the respondents were still not aware of arsenic contamination. It was observed by the author that awareness raising campaigns in the mass-media on arsenicosis has significantly reduced in 2011–2012, as compared to 2003–2004. So, it is natural for those households not having arsenicosis patients in their households to remain unaware or poorly aware of the toxicity of arsenic and what they should do to save themselves from arsenicosis.

9.3.2   Willingness to Pay per Month

Willingness to pay often measures the level of awareness among the respondents. People in Bangladesh are not used to *paying* for the water they use. Prior to arsenic poisoning they used tube well water almost free of cost. Therefore, we find that just as the level of awareness had risen in 2005 as compared to 2000, their willingness to pay per month had also risen (a mean of Taka 36.32 to Taka 59.71). Whereas almost 30% (n = 36) of the respondents refused to pay any money for safe water in 2000, there was only one respondent (0.6%) who was not willing to pay any money to get safe water. Furthermore, there were more who were willing to pay above Taka 50 in 2005 (11.9%, n = 21) than in 2000 (3.2%; n = 4). During 2011–2012, 56.2% of the respondents were willing to pay for getting safe water. The rest (43.7%) said that they earned very little and had loans to repay for which they were not able to pay any money to get safe water. When they were requested to suppose that they had already repaid the loan, and asked in that case how much they were willing to pay, one of the respondents said he was not willing to pay any money to get water even after becoming loan-free. In Table 9.5 we find that more than 27% of the respondents were not willing to pay at all to get safe water, 68.5% wanted to pay only up to Taka 50 per month and only 4% were willing to pay up to Taka 100. It is thus observed that willingness to pay reduced after the situation regarding arsenicosis patients had improved in Tarananagar (in 2012). The reason for this could be that they were already enjoying the facility of a safe water source near their home and so did not feel it necessary to pay any money to get the same.

Table 9.6 Willingness to donate for the establishment of safe water in 2000, 2005 and 2012

Willingness to donate (Taka)	2000		2005		2012	
	No. of respondents	Percentage	No. of respondents	Percentage	No. of respondents	Percentage
00	55	57.9	0	00.0	47	18.7
00–500	30	31.6	160	91.4	197	78.5
500–1000	10	10.5	6	3.4	6	2.4
1000–2000	00	00.0	5	2.9	0	0.0
2000–5000	00	00.0	4	2.3	1	0.4
Total	95	100.0	175	100.0	251	100

9.3.3   Willingness to Donate

Respondents were asked how much they were willing to donate for establishment of community based safe water sources. Just as we have observed in the case of willingness to pay, peoples' willingness to donate has increased significantly in 2005 as compared to 2000. Whereas almost 58% of the respondents refused to donate any money in 2000, all respondents in 2005 have agreed to donate at least Taka 500 for the establishment of a safe water option. The highest amount of money respondents in 2000 were willing to donate was Taka 1000; in 2005 they were willing to donate up to Taka 5000. It was observed that like willingness to pay per month, willingness to donate to establish a safe water system also reduced in 2012. Almost 19% of the respondents were not willing to donate at all. Among those who were willing to donate, 78.5% were willing to give only up to Taka 500 (Table 9.6).

9.3.4   Willingness to Walk

It is assumed that for each person five liters of water is required for the purposes of drinking and cooking. If the distance of the water source is far from home compound, it is quite difficult for people to carry so much water. In Bangladesh, procuring necessary water for all the family members is traditionally considered 'women's work.' Therefore, the female heads of household were asked how much distance they were willing to walk to bring safe water (see Table 9.7). In 2000 almost 17% of the respondents (n = 16) were not willing to walk at all. However, in 2005 only 6.9% of the respondents (n = 12) were unwilling to walk at all. Most surprisingly almost 75% of the respondents were willing to walk a distance of even 20 min to bring water in 2005. As there were more safe water sources in Taranagar now (2012) than in 2005, the female respondents' willingness to walk to bring water also reduced. In 2012, 33.5% of the respondents were not willing to walk at all to fetch safe water. Most of them, however, (58.2%) were willing to walk a 5 min' distance, while 6% were willing to walk a distance of up to10 min. The reason for their unwillingness to walk was that 48.2% of the respondents were using a water source within their

Table 9.7 Women's willingness to walk to bring safe water in 2000, 2005 and 2012

Willingness to walk (Minutes)	2000		2005		2012	
	No. of respondents	Percentage	No. of respondents	Percentage	No. of respondents	Percentage
00	16	16.8	12	6.9	84	33.5
00–05	69	72.6	8	4.6	146	58.2
5–10	10	10.5	3	1.7	15	6.0
10–15	00	00.0	21	12.0	4	1.6
15–20	00	00.0	131	74.9	2	0.8
Total	95	100.0	175	100.0	251	100.0

Table 9.8 Respondents' first choice regarding alternative option in 2000, 2005 and 2012

Options	2000		2005		2012	
	No. of respondents	Percentage	No. of respondents	Percentage	No. of respondents	Percentage
Tap water at home	95	100.0	55	31.43	247	98.4
Tap water in walking distance	00	00	113	64.57	3	1.2
Purified dug well/pond water	00	00	5	2.86	0	0
Filters	00	00.0	2	1.14	0	0
Dug well/ ring well	00	00	00	00	1	0.4
Total	95	100.0	175	100.0	251	100.0

home compound, and hence not habituated to walk a distance to bring water. As the ratio of households having patients had reduced significantly, their willingness to walk also reduced. Moreover, now (2012) there were more safe water sources near the homes of the inhabitants of Taranagar. So, the necessity to walk long distances also had reduced.

9.3.5   Preference of the Respondents Regarding Alternative Safe Water Options

In 2000 all respondents chose tap water at home as their first preference in the hierarchy of safe water options. In 2005, respondents have become more realistic. They knew that it would be very costly to get tap water in their home compound, so as first preference most respondents have chosen tap water at a close distance. However, in 2012 respondents reiterated their preference (98.4%) for tap water at home (see Table 9.8). Only 1.2% of them opted for tap water within a close distance as their first preference. The author spoke to some of the women folk of Taranagar in 2012, who said that they were quite content to collect water from ring wells close to their home compounds.

In order to provide tap water, there could be two options regarding the source. One is deep tube well; the other is the big beel (depression). The cost however, is very high; and in this regard, the residents of Taranagar alone would not be able to bear the cost of establishment. As we have seen the respondents were not very much interested to pay for getting safe water. Moreover, neither of the two alternatives was an acceptable feasible option to the respondents. Deep tube wells have proved to be non-feasible in Taranagar, and it would be difficult to have consent of all concerned to use the big depression of the village (which was being used mainly for fish culture) as the source of drinking water, which would mean that all other uses of the depression would have to stop. The other options were of a smaller magnitude, e.g., pond sand filter, rain-water harvesting and filtration of (contaminated) tube well water.

9.4 Efforts for Arsenic-Mitigation in Taranagar

The Department of Public Health Engineering (DPHE) is the government institution responsible for providing safe water to the arsenic affected people. With an attempt to provide safe water, the DPHE of Meherpur and some NGOs built 27 ring wells for the arsenic-affected people in the village. However, only 18 were functioning at the time of the survey in 2012. These ring wells were 30–40 f deep. There were three dug wells (80–90 f deep) in use at the time of the survey. However, these failed to supply the badly hit people with sufficient safe water. There have been accusations that the influential Chairman/members of the Union Parishad (the grass root level of local government) tried to locate the ring wells within their own areas, depriving those who needed it more urgently (like the residents of Taranagar). Efforts have been made to supply deep tube well water to Tarnagar people. However, water from deep layers had been found to be highly contaminated with arsenic in Taranangar. An NGO named Manab Shakti Unnayan Kendra (Human Resource Development Centre) distributed some SONO filters, most of which were being used by the villagers. SONO filter was approved by the government as safe to use. Another NGO named KARITAS distributed some arsenic filters. Usually one filter was used by one family. However, the filters required regular maintenance, which was often not properly done by the villagers. In 2012 BRAC distributed about 280 SONO filters to the villagers of Taranagar. Eighty of them were donated by an American citizen. BRAC distributed these filters buying the filters from Manab Shakti Unnayan Kendra. Five households in Taranagar had rainwater purification systems in their houses, but as the area experienced less than 1500 mm of rain per year on an average, the procured rainwater was not sufficient for the whole year. Five NGOs were working for arsenic related matters in Taranagar in 2012. These NGOs were Bangladesh Rural Advancement Committee (BRAC), KARITAS,

Table 9.9   Respondents' opinion on the government's efforts in arsenic-mitigation

Respondents' opinion	2000		2005		2012	
	No. of respondents	Percentage	No. of respondents	Percentage	No. of respondents	Percentage
No	75	78.9	92	52.6	197	78.5
Yes	20	21.1	83	47.4	54	21.5
Total	95	100.0	175	100.0	251	100.0

Table 9.10   Respondents' opinion on NGOs' efforts in arsenic-mitigation

Respondents' opinion	2000		2005		2012	
	No. of respondents	Percentage	No. of respondents	Percentage	No. of respondents	Percentage
No	21	22.1	44	25.1	110	43.8
Yes	74	77.9	131	74.9	141	56.2
Total	95	100.0	175	100.0	251	100.0

Social Development, Heed Bangladesh and Manab Shakti Unnayan Kendra. These NGOs had been working to raise awareness with regard to arsenic poisoning, establishment and distribution of safe water sources/instruments, distribution of vitamins to patients and in some cases distribution of cash money. It was learnt from the survey that during winter (November to February), the villagers hardly drank filter water, as the water was very cold. During this period they preferred to drink tube well water or ring well/dug well water as these sources naturally provided water that is warmer than ground temperature. It has been revealed in the survey results of 2011–2012 that there were various reasons for which the villagers preferred to use dug well and ring well water all year round. It was also observed in the village that although five NGOs and the DPHE had taken initiatives to combat the arsenic problem in Taranagar, no initiative was taken to monitor the existing sources of drinking/cooking water being used by the villagers. Moreover, there was no community organization in the village to integrate all the efforts, which in the long run prevented these projects from becoming cost-effective. Also as there was no central authority to monitor the distribution system of the arsenic-safe water sources, the situation was such that a household enjoyed two or more sources adjacent to their homes, while others had to walk long distances to procure safe water and often continued to drink from unsafe tube wells for that reason.

Many of the respondents did not recognize that the government/NGOs have done something to mitigate the problem. The DPHE claimed that under different projects all the tube wells were screened, and some ring wells have been excavated. The NGOs claimed that they were engaged in putting red mark on contaminated tube wells and green mark on safe ones. The respondents' opinion regarding the government/NGO efforts is shown in Tables 9.9 and 9.10. Table 9.9 shows that the respondents were more aware of government efforts in 2005 than in 2000. More than 50% of the respondents in 2005 still said that they did not know about government efforts in arsenic mitigation. This proportion was much greater in 2012 (over 78%). In case of NGO efforts however, lower percentage of respondents recognized arsenic-mitigation efforts in 2005 than in 2000. In 2012 almost 44% of the respondents could not recognize any visible effort by NGOs to mitigate the arsenic problem in Taranagar.

Chapter 10
Conclusions

Abstract This chapter presents a summary of the research findings that have been discussed in the whole book and draws a conclusion by making some policy suggestions, specifically for villages like Taranagar. The author observes that the mitigation efforts in the village lack an integrated approach. Referring to the suggestions made by Kunikane (An integrated Approach to Arsenic Mitigation in Bangladesh, http://www.who.int/household_water/resources/kunikane.pdf, 2009), the author recommends that functioning of an efficient arsenic mitigation committee nominated by the villagers with financial support of the wealthy people was needed to ensure that all residents of Taranagar drink and cook food with safe water.

10.1 Summary of Findings

The present study has examined socio-economic and arsenic related situation mainly in three badly arsenic affected villages in Meherpur District (Bagoan, Dholmari and Taranagar), Bangladesh through a sample survey, developed a multi-objective mixed-integer Pareto optimization model in order to find the optimal option for the residents of Tarangar, carried out simulations to test its validity, pointed out the mechanisms through which it might be possible to convince the people to adopt a safe water option as a substitute for tube wells, and chalked out a system to provide safe water to the arsenic affected people of Taranagar. At the outset of the research, an e-mail survey was carried out among experts to know about their views on the cause of the problem and how it could be solved. Among the respondent experts, 58.33% supported the Reduction of Iron Oxyhydroxide hypothesis, and 33.33% supported the Oxidation of Pyrite hypothesis and 75% of the respondents believed that over-extraction of ground water had some relationship with the contamination of arsenic in ground water in Bangladesh. Most of the experts refrained from giving any specific answer to the question on how the problem should be solved.

The inhabitants of the three villages initially under study in the year 2000 (Bagoan, Taranagar and Dholmari) were mainly farmers (engaged in agriculture, cultivating their own land) and agricultural laborers. The monthly income of the respondents varied from 1500 Taka (US\$ 75) to above 10,000 Taka (US\$ 2000). However, 55%–60% of the inhabitants earned 2000 Taka or less. The average schooling years of households was between 3.47 and 4.2 years. Taranagar was the village in which, the highest percentage of tube wells were contaminated (80.64%)

© Springer Japan KK 2017

W. Akmam, *Arsenic Mitigation in Rural Bangladesh*, New Frontiers in Regional
Science: Asian Perspectives 16, https://doi.org/10.1007/978-4-431-55154-6_10

among the three villages. Percentage of households having arsenicosis was also the highest in this village (72.72%). Most of the womenfolk in all the three villages were willing to walk for 5 min to bring safe water, and most of the heads of households were willing to pay only US$1.00 per month for that purpose in the year 2000.

Cross tables had been prepared to apply Chi-square test and calculate Pearson's Correlation Coefficients in order to determine the association/correlation between variables. In Taranagar, the only two correlated variables were income and willingness to pay. In Bagoan, land-patient, income-patient, education-patient, willingness to pay-patient, donation-patient were found to be significantly associated and education-willingness to pay, income-willingness to pay, income-donation and income-willingness to walk were found to be significantly correlated. In Dholmari, the variables land and patient were significantly associated; education-donation and income-donation were significantly correlated.

Using the data found through the social survey, and some other secondary data on the different safe water options, a multi-objective mixed-integer model based on Pareto optimality was developed. Twelve different types of safe water options were considered, which included deep tube wells, surface water filters, surface water treatment plants, dug wells, rainwater harvesting systems and different types of filtering methods to remove arsenic. The author constituted a set of structural equations in order to supply safe water for the people living in Taranagar village, which was heavily dependent on the optimal solution to the model. The optimization considered not only an optimum selection among the set of safe options, but also the optimality of the selected safe option all over the village. The latter optimality was dependent on multi-objects, namely the number of people who could get safe water on the specified standard, the total cost necessary for construction and maintenance, environmental friendliness of the selected safe options, amount of time required to be spent by the villagers to get the safe water, geographical feasibility, ease in operation etc. As mentioned earlier, in this model, a Pareto optimization approach was used, an approach in which the multi-objects were to be maximized (or minimized) being subject to other multi-objects to be retained at the specific level jointly with several structural equations.

Simulation results (based on data collected in 2000) showed that the most efficient locus of the trade-off was to be found in the establishment of 18 dug wells in specific zones (see Fig. 6.2) and providing each household with a surface water filter, with which they could filter the dug well water before drinking. The inhabitants of Taranagar would have to walk about 2 mins' distance in order to bring dug well water from the closest facility from their own houses. This option will satisfy all constraints except the WHO standard for arsenic (Case 1). If we take into account all the constraints (Case 2), the optimal option would be establishment of 15 community based Activated Alumina Filters in specific zones (see Fig. 6.3). However, the total cost will increase more than fourfold. The trade-off point for cost and distance for Case 1 was at an average distance of 58 m, and for Case 2 the trade-off point was at an average distance of 67 m. Therefore, as was revealed through the simulation results, the model built was valid and could be applied for other affected areas of Bangladesh, using data for respective areas. The multi-objective mixed-integer programming model developed in this study was essential for the subsystems, which constituted the collective decision making system based on informed consent. In order for the affected people to make a decision on the safe water options, a mere collection of speculated cost information on the available options was insufficient. A profile of efficient options along which they might be able to make a trade-off among several competing objectives would make a big difference for them, especially as the common people of Taranagar village were not able to identify or imagine the efficient profile by solving optimal assignment of options over the study area. While the model was of a mixed-integer form, step-wise approximations by hyperbolic functions were proved effective in practice to get an optimal solution to the model.

The author has chosen Taranagar as the village as the study area to carry out longitudinal research on arsenic mitigation. Survey results of Taranagar village in 2005 showed that the situation of the people had worsened significantly as compared to the 2000. Their average monthly income had lowered. However, their level of awareness had risen. The most alarming finding was that almost all households had patients, but almost half of them continued to drink water from contaminated tube wells. A few ring wells had been dug in order to provide safe water, but the water of the wells contained bad odor. In 2012 the percentage of households with arsenicosis patients had come down to 55.4%, meaning that awareness of arsenic had increased significantly and some interventions in combating the arsenic-related situation had taken place. However, only 49% of the respondents interviewed in 2012 were sure that they were drinking from safe sources, and 6.4% did not know whether their sources were safe or not. More than 42% of the respondents continued to drink water from tube wells, 90% of which were contaminated.

10.2 Policy Suggestions

The author's observation is that the mitigation efforts going on in Taranagar lacks an integrated perspective. In his integrated effort, Kunikane (2009) included six main approaches, e.g., surveillance of arsenic contamination in groundwater; raising community awareness; community organization; providing a supply of safe water with training to maintain the sources; testing water quality of the sources; and providing health care to arsenicosis patients. If we evaluate the situation in Taranagar based on these six approaches, we find that none have been followed properly. Surveillance of arsenic contamination in ground water in Taranagar was almost totally absent. The screening of tube wells was done only once in the village during 1999–2000 period. Since then no such screening had taken place. The practice observed in 2012 was that the person who was interested to check whether his tube well was spewing safe water or not would collect a sample and submit it at the DPHE office to test it, after paying the required fee (Taka 500 per sample). Often paying this fee became impossible, for which tube wells (even new ones) were not checked. Awareness raising campaigns have almost stopped nationally, and specifically in Taranagar as well. Therefore, we have seen a discouraging situation regarding awareness of the people in the village in 2012. There existed no community organization in Taranagar that could be given the responsibility for integrating the mitigation works that took place in Taranagar. As far as the author has learnt on the issue, the inhabitants did not receive any training to maintain the safe water sources, for which the sources that had become out of order remained unused for a long time. Proper monitoring of the water quality of the existing safe water sources was also not ensured. The Meherpur Medical Hospital provided some treatment for the arsenicosis patients. Sometimes doctors/nurses visited the village and prescribed and distributed necessary medicines to the patients. This helped to reduce the number of households having patients in Taranagar.

The author believes that an efficient arsenic mitigation committee nominated by the villagers, who would be dedicated and held responsible for integrating all arsenic related affairs, could provide a solution to all the arsenic related problems in Taranagar and provide safe water sources for everyone at a reasonable distance from their homes. At present, the DPHE and NGOs working in the village are distributing filters and establishing safe water sources, without appointing specific persons who would ensure that the sources remain safe and function properly.

It must also be brought under consideration that the arsenic mitigation committee in Taranagar would require some funds to accomplish their tasks in a comprehensive manner. The solvent/capable villagers could pay small amounts regularly to help this committee to function properly; this would also ensure that these donors will remain vigilant over the committee members' performance.

10.3   Conclusion

In conclusion, it may be stated that the government and NGOs have contributed sufficiently to provide safe water to the arsenic hit people of Taranagar. However, some people in the village were still drinking water from arsenic-contaminated tube wells due to ignorance and lack of proper monitoring and management. A functioning, efficient and dedicated arsenic mitigation committee in Taranagar could make sure that all the arsenic-hit people in the village get equitable access to safe water. We must bear in mind that honesty, sincerity and integrity of all the stakeholders is essential in combating this formidable challenge.

Appendices

Appendix 1

Elaborations of Abbreviations

ASY	Average Schooling Years
BAMWSP	Bangladesh Arsenic Mitigation and Water Supply Project
BGS	British Geological Survey
BRAC	Bangladesh Rural Advancement Committee
DCH	Dhaka Community Hospital
Don.	Willingness to Donate at the Initial Stage
DPHE	Department of Public Health Engineering, (Bangladesh)
Edu.	Schooling of the Respondent
JICA	Japan International Cooperation Agency
JU	Jadavpur University
NGO	Non-Government Organization
NRA	National Research Council
NSA	No Significant Association
NSC	No Significant Correlation
ODA	Official Development Assistance
Pa.	Arsenic Patient
RPMU	Regional Project Management Unit
SA	Significant Association
SC	Significant Correlation
TAG	Technical Advisory Group
UNICEF	United Nations Children's Fund
WE	Women's Education (Schooling Level of the Female Respondent)
WHO	World Health Organization
WP	Willingness to Pay
WW	Willingness to Walk

© Springer Japan KK 2017
W. Akmam, *Arsenic Mitigation in Rural Bangladesh*, New Frontiers in Regional
Science: Asian Perspectives 16, https://doi.org/10.1007/978-4-431-55154-6

Appendix 2

Map 1 Bangladesh (Districts) (Source: http://www.lged.gov.bd/images/Bangladesh.jpg. Accessed 11 November 2017)

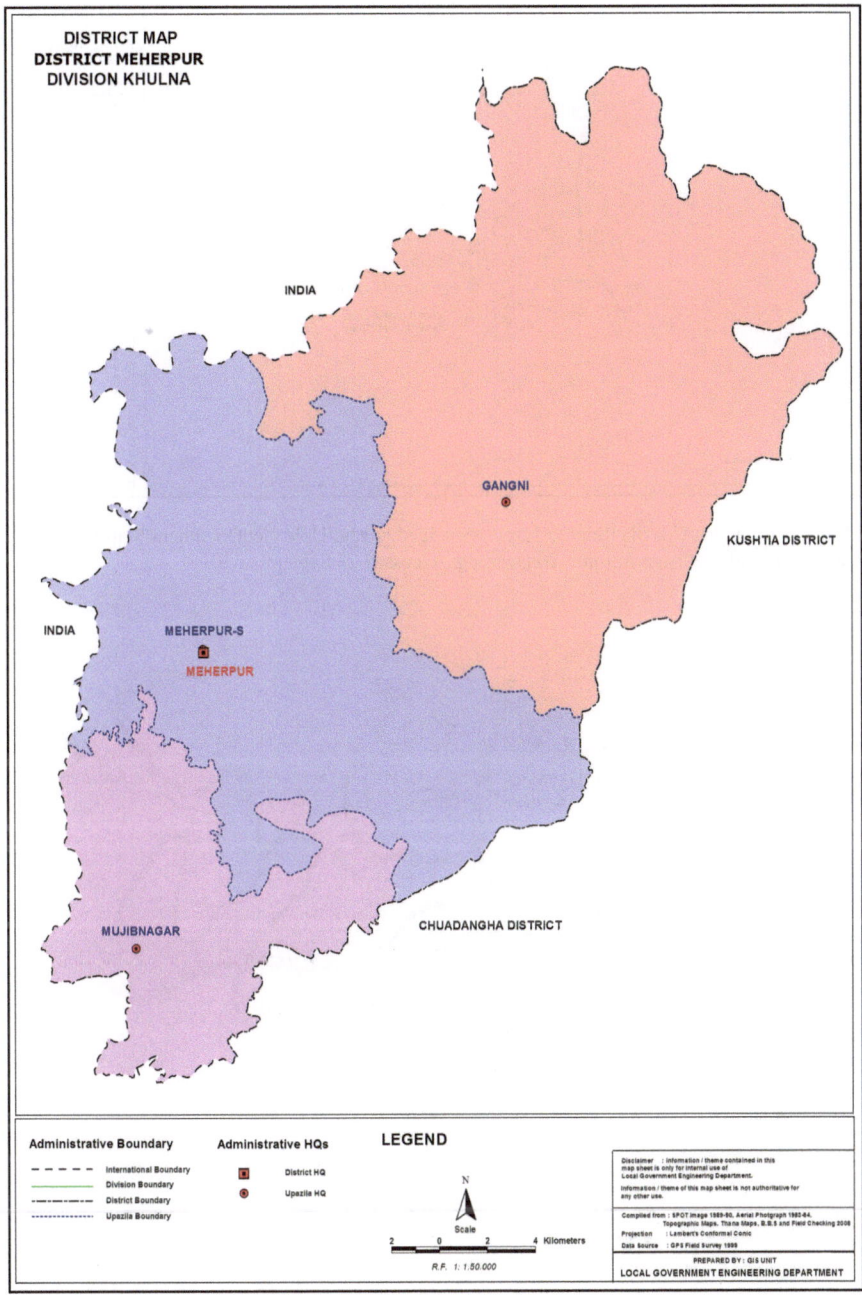

Map 2 Meherpur District (Source: http://www.lged.gov.bd/DistrictHome.aspx?districtID=43&fromDistrictpage=2. Accessed 11 November 2015, 2017)

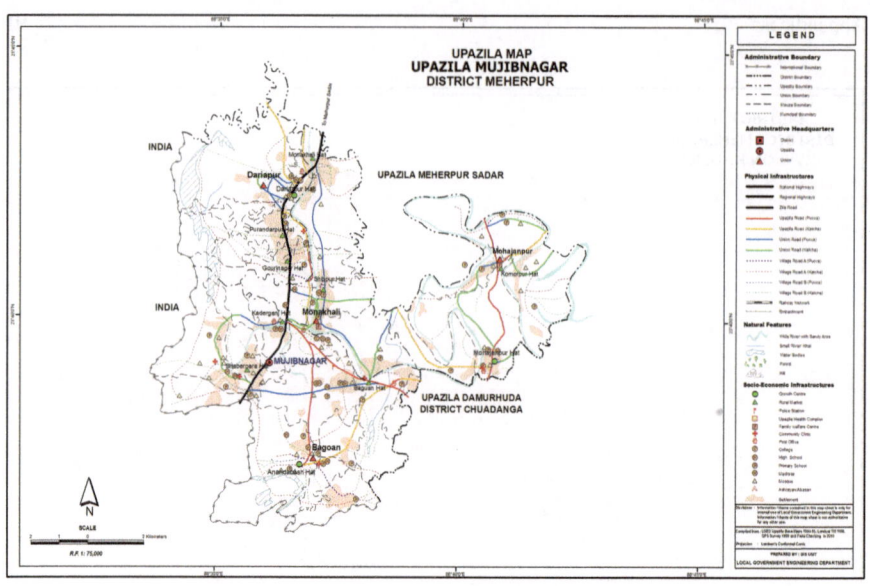

Map 3 Mujibnagar Upazila (Source: http://www.lged.gov.bd/UploadedDocument/Map/KHULNA/meherpur/mujib%20nagar/mujib%20nagar.jpg. Accessed 28 September 2015)

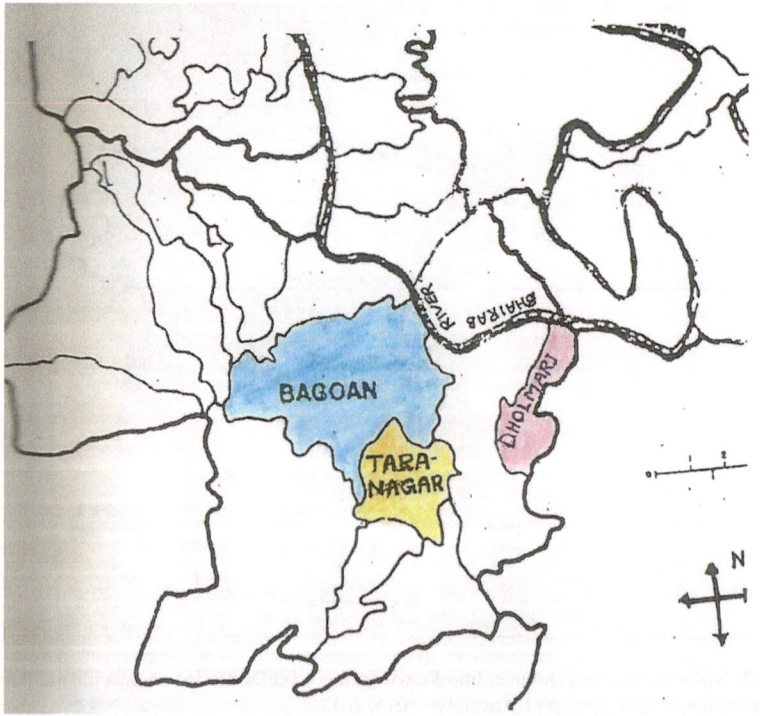

Map 4 The Study Villages (Bagoan, Taranagar and Dholmari) Source: Government of the People's Republic of Bangladesh (1989). *Small Area Atlas of Bangladesh: Mauzas and Mahallas of Kushtia District*. Dhaka: Bangladesh Bureau of Statistics (The original map was modified by the author)

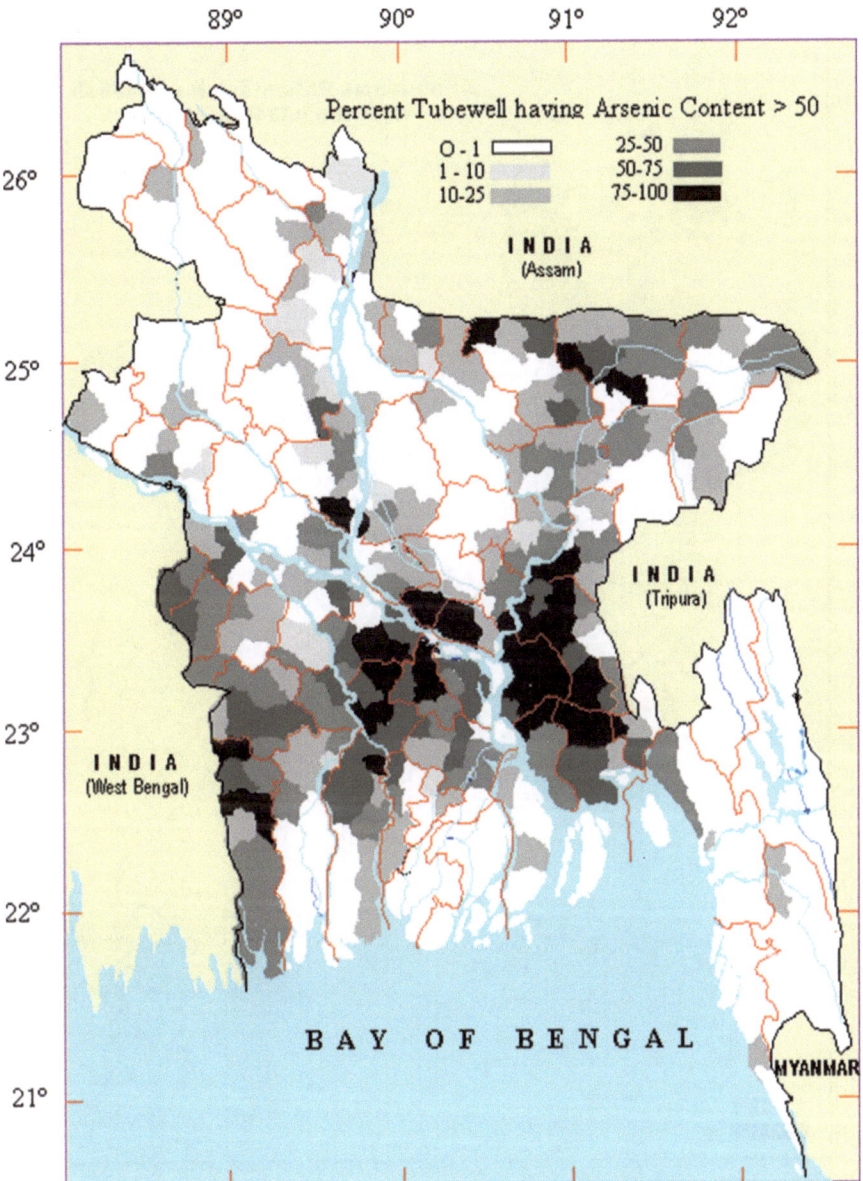

Map 5 Bangladesh (Intensity of Arsenic Poisoning in Terms of Affected Tube Wells) (http://users.physics.harvard.edu/~wilson/arsenic/conferences/Feroze_Ahmed/Sec_2.ht. Accessed on 29 September 2015)

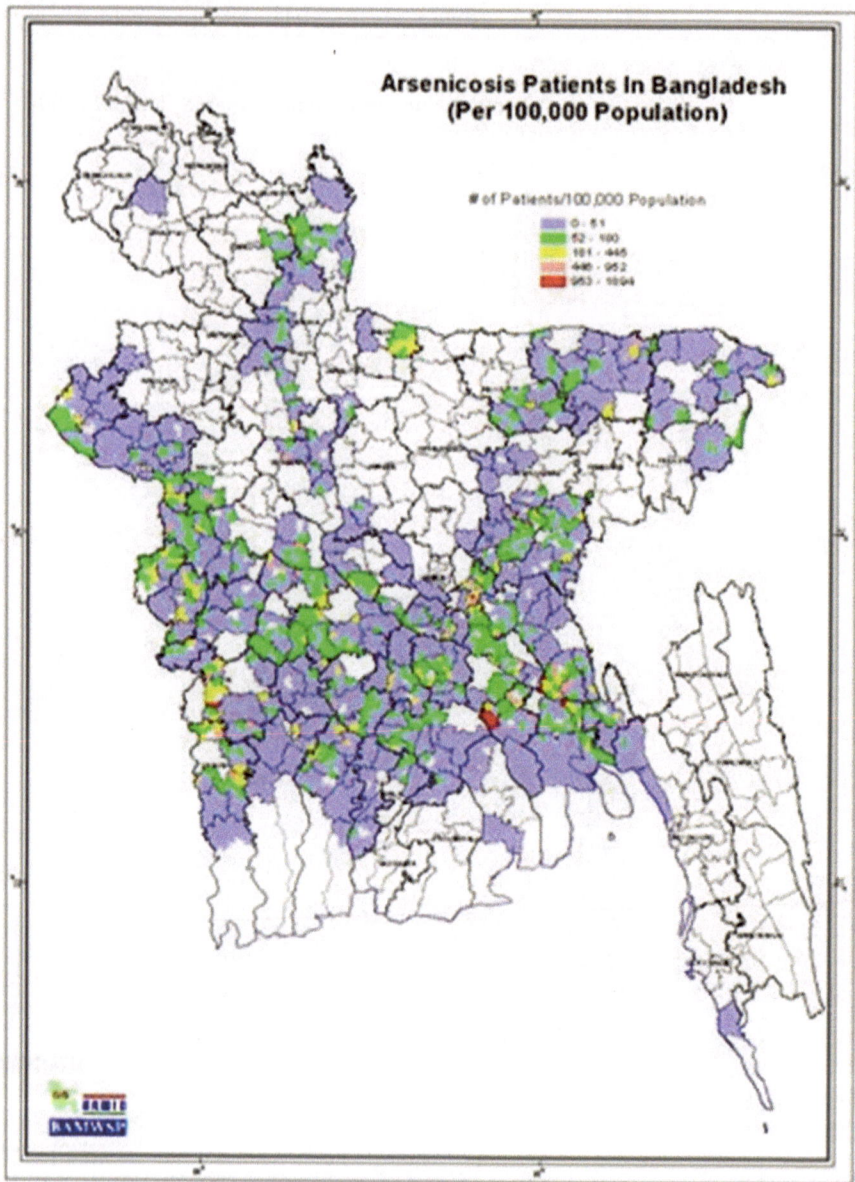

Map 6 Arsenicosis Patients in Bangladesh (Source: (Department of Public Health Engineering, Bangladesh) http://www.dphe.gov.bd/index.php?option=com_content&view=article&id=96&Ite mid=104. Accessed 13 October 2016)

Appendix 3

Appendix 3.1

Pictures of Arsenicosis Patients

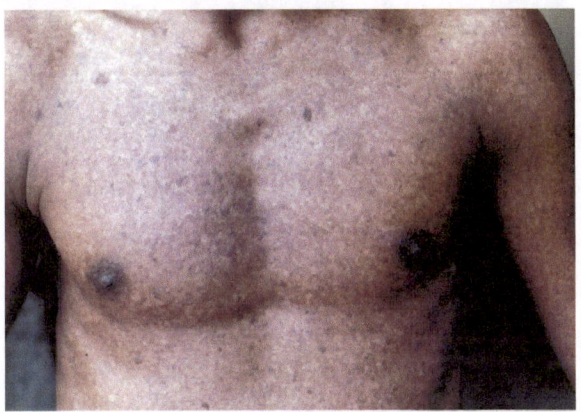

Chest of an arsenicosis patient. Photo source: NGO Forum, Bangladesh

Hands of a Keratosis patient. Photo source: NGO Forum, Bangladesh

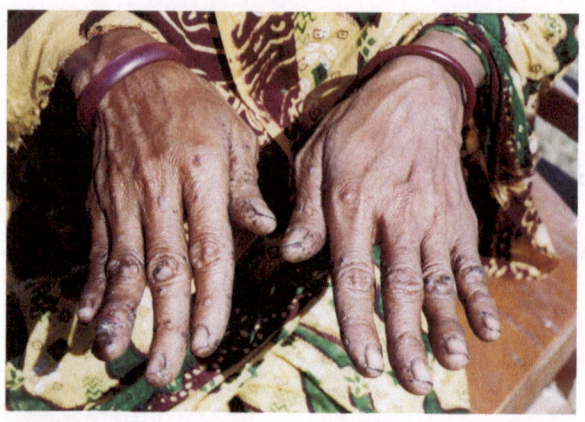

Hands of an arsenicosis patient. Photo source: NGO Forum, Bangladesh

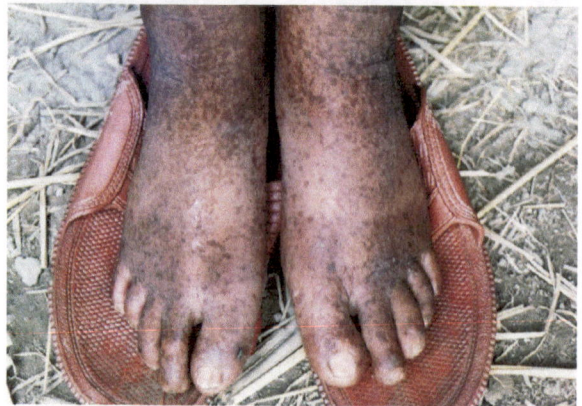

Feet of an arsenicosis patient. Photo source: NGO Forum, Bangladesh

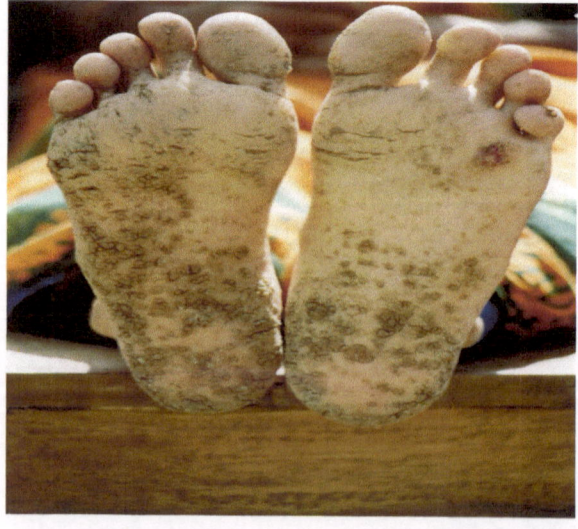

Keratosis (sole) of an arsenicosis patient. Photo source: NGO Forum, Bangladesh)

Appendix 3.2

Pictures of Some Arsenic-Safe Sources of Water

Picture of a Pipeline Water Supply System. Photo source: NGO Forum, Bangladesh

Picture of a Pond Sand Filter: Photo source: NGO Forum, Bangladesh

Picture of a Rainwater Harvesting System: Photo source: NGO Forum, Bangladesh

Appendix 3.3

Pictures of Taranagar

Picture 1 The Researcher at the Site of Newly Built (April 2011) Dug Well at Taranagar Village (East Section)

Picture 2 The Researcher at the Site of Newly Built Ring Well at Taranagar Village (North Section)

Picture 3 The Researcher
at the Site of Newly Built
Ring Well at Taranagar
Village (Central Section)

Picture 4 The Researcher Talking with the Villagers

Picture 5 The Researcher at the Site of a Newly Built Ring Well at Taranagar Village Talking with the Key Informant Mr. Kiamat Ali

Picture 6 The Researcher at the Site of Newly Built Ring Well at Taranagar Village (Central Section) Talking with the Women of Taranagar Village who Collect Drinking Water

Picture 7 A Villager Seems Happy with the New Ring Well at Taranagar Village (Central Section)

Picture 8 The Researcher at the Site of Newly Built Ring Well at Taranagar Village (North Section) Talking with the Women

Picture 9 A Woman Collecting Water from the Newly Built Ring Well at North Section

References

Published Materials

Adhikarya, R. 1989. Strategic extension campaigns on rat control in Bangladesh. In *Public communication campaign*, 2nd ed, ed. R.E. Rice and C.K. Atkin, 230–232. Beverly Hills/London/New Delhi: Sage Publications.

Ahmed, H.S. 2000. *Combating a deadly menace: Early experiences with a community-based arsenic mitigation project in Bangladesh—June 1999–June 2000*. Dhaka: BRAC.

Ahmad, N. 1976. *Economic geography of Bangladesh*. New Delhi: Vikas Publishing House.

Ahmad, S.A., M.H. Salim Ullah Sayed, S. Barua, M.H. Khan, M.H. Faruquee, Abdul Jalil, et al. 2001. Arsenic in drinking water and pregnancy outcomes. *Environmental Health Perspectives.* 109 (6): 629–631.

Ahmad, J., B. Goldar, and S. Misra. 2006. Rural communities' preferences for arsenic mitigation options in Bangladesh. *Journal of Water Health* 4 (4): 463–477.

Ahmed, H.S. 2000. *Combating a Deadly Menace: Early Experiences with a Community-Based Arsenic Mitigation Project in Bangladesh—June1999–June 2000*. Dhaka: BRAC.

Ahmed, M.F. 2002. *Alternative Water Supply Options for Arsenic Affected Areas in Bangladesh. International Workshop on Arsenic Mitigation in Bangladesh*. Dhaka: Centre for Water Supply and Waste Management.

Ahmed, K.M. 2003. Arsenic contamination in groundwater and a review of the situation in Bangladesh. In *Groundwater Resources and Development in Bangladesh: Background to the Arsenic Crisis, Agricultural Potential and the Environment*, ed. A.A. Rahman and P. Ravenscroft, 395–417. Dhaka: UPL.

Akmam, W., and M.F. Islam. 2007a. Arsenic contamination in ground water in Bangladesh: A study on Taranagar village. *Studies in Regional Science* 37 (3): 829–840.

———. 2007b. Factors affecting awareness regarding arsenic poisoning in Bangladesh. *Eubios Journal of Asian and International Bioethics* 17 (2): 54–62.

Akmam, W., and F. Islam. 2008. Taranagar from 2000 to 2005: Changes in arsenic-related situation in a badly affected village in Meherpur, Bangladesh. *Studies in Regional Science* 38 (1): 17–31.

Akmam, W., S.M.H. Marshad, and Md Fakrul Islam. 2007. Arsenic contamination in ground water in Bangladesh: Nature of awareness in an arsenic-prone area. *Eubios Journal of Asian and International Bioethics* 17 (6): 162–166.

Alves, M.J., and J. Climaco. 2000. An interactive reference point approach for multi-objective mixed integer programming using branch-and-bound. *European Journal of Operational Research* 124: 478–494.

© Springer Japan KK 2017
W. Akmam, *Arsenic Mitigation in Rural Bangladesh*, New Frontiers in Regional
Science: Asian Perspectives 16, https://doi.org/10.1007/978-4-431-55154-6

Anwar, J. 2000. *Arsenic Poisoning in Bangladesh: The end of a Civilization?* Dhaka: Palash Media and Publishers.

As-contamination Research Group, Niigata University. 2000. Constituent minerals of drilling samples at Samta, in Bangladesh and their characteristics for arsenic contaminated groundwater. *Earth Science (Chikyu Kagaku)* 54: 94–104.

Atkin, Charles K., and Hendrika W.J. Meischke. 1989. Family planning communications campaigns in developing countries. In *Public Communication Campaign*, ed. Ronald E. Rice and Charles K. Atkin, 2nd ed., 227–229. Beverly Hills/London/New Delhi: Sage.

Bae, M., C. Watanabe, T. Inaoka, M. Sekiyama, N. Sudo, M.H. Bokul, and R. Ohtsuka. 2002. Arsenic in cooked rice in Bangladesh. *The Lancet* 360 (3948): 1839–1840.

Barendra Advancement Integrated Committee (BAIC). 2004. *BAIC Annual Report 2003–2004.* Chapai Nawabganj: BAIC.

Bates, F.L., and C.C. Harvey. 1975. *The structure of social systems.* New York: Gardner Press.

Berrien, F.K. 1968. *General and social systems.* New Brunswick: Rutgers.

Brinkel, J., M.M.H. Khan, and A. Kraemer. 2009. A systematic review of arsenic exposure and its social and mental health effects with special reference to Bangladesh. *International Journal of Environmental Research and Public Health* 6 (5): 1609–1619. doi:10.3390/ijerph6051609.

British Geological Survey. 2009. Implications of the present study for the arsenic mitigation strategy. http://www.bgs.ac.uk/arsenic/bphase1/B_impl.htm. Accessed 13 Jan 2009.

Caldwell, B.K., et al. 2006. Access to drinking-water and arsenicosis in Bangladesh. *Journal of Health Population and Nutrition* 24 (3): 336–345.

Chang, C. 2000. An efficient linearization approach for mixed-integer problems. *European Journal of Operational Research* 123: 652–659.

Charnes, A., and W.W. Cooper. 1961. *Management Models and Industrial Applications of Linear Programming.* New York: Wiley.

Chen, C.J., Y.M. Hsueh, H.Y. Chiou, Y.H. Hsu, S.Y. Chen, S.F. Horng, K.F. Liaw, and M.M. Wu. 1997. Human carcinogenicity of inorganic arsenic. In *Arsenic Exposure and Health Effects*, ed. C.O. Abernathy, R.L. Calderon, and W.R. Chappell. London: Chapman and Hall.

Chen, Y., et al. 2007. Reduction in urinary arsenic levels in response to arsenic mitigation efforts in Araihazar, Bangladesh. *Environmental Health Perspectives.* 115 (6): 917–923.

Chowdhury, S. 2000. Arsenic Crawling death. *The Courier,* February 4, 2000.

Chowdhury, S. N. 1998. Safe Drinking Water for All: Containing Arsenic Contamination: An NGO Initiative. *The Daily Star* 16 July, 1998.

Chowdhury, T.R., G.K. Basu, B.K. Mandal, B.K. Biswas, G. Samanta, U.K. Chowdhury, C.R. Chanda, et al. 1999a. Brief communication: arsenic poisoning in the Ganges delta -- the natural contamination of drinking water by arsenic needs to be urgently addressed. *Nature* 401: 545–546.

Chakraborti, D., S. Kabir, and S. Roy. 1999. Groundwater arsenic contamination and suffering of people in Bangladesh. In *Arsenic exposure and health effects*, ed. W.R. Chappell, C.O. Abernathy, and R.L. Calderon, 165–182. Oxford: Elsevier.

Crema, A. 2000. An algorithm for the multi-parametric 0-1 integer linear programming problem relative to the objective function. *European Journal of Operational Research* 125: 18–24.

Das, H.K., A.S.M.F. Mallick, and P.K. Sengupta. 2003. Health effects of arsenic contamination in groundwater. In *Groundwater Resources and Development in Bangladesh: Background to the Arsenic Crisis, Agricultural Potential and the Environment*, ed. A.A. Rahman and P. Ravenscroft, 418–433. Dhaka: UPL.

Dhaka Community Hospital. 2002. *Arsenic problem.* http://www/dchtrust.org. Accessed on 12 Apr 2002.

Edmunds, W.M., K.M. Ahmed, and P.G. Whitehead 2015. *A review of arsenic and its impacts in groundwater of the Ganges–Brahmaputra–Meghna delta*, Bangladesh. *Environmental Science: Processes & Impacts.* doi:10.1039/c4em00673a. http://www.espa.ac.uk/results/publications/review-arsenic-and-its-impacts-groundwater-ganges-brahmaputra-meghna-delta. Accessed 26 Aug 2015.

Fagence, M. 1977. *Citizen participation in planning.* Sydney: Pergamon Press.

Fazal, M.A., T. Kawachi, and E. Ichion. 2001a. Extent and severity of groundwater arsenic contamination in Bangladesh. *Water International* 26 (3): 370–379.

————. 2001b. Validity of the latest research findings on causes of groundwater arsenic contamination in Bangladesh. *Water International* 26 (3): 380–389.

Flanagan, S.V., R.B. Johnston, and Y. Zheng. 2012. Arsenic in tube well water in Bangladesh: Health and economic impacts and implications for arsenic mitigation. *Bulletin of the World Health Organization* 9: 839–846. doi:10.2471/BLT.11.101253. Accessed 22 Aug 2015.

Friend, J.K., and W.N. Jessop. 1977. *Local government and strategic choice: An operational research approach to the process of public planning*, 2nd ed. New York: Pergamon Press.

Geoffrion, A.M. 1967. Solving bicriterion mathematical programs. *Operations Research* 15: 39–54.

Geoffrion, A.M., J.S. Dyer, and A. Feinberg. 1972. An interactive approach for multi-criteria optimization, with an application to the operation of an academic department. *Management Science* 19: 357–368.

Government of the People's Republic of Bangladesh. 1989. *Small area atlas of Bangladesh: Mauzas and Mahallas of Kushtia District*. Dhaka: Bangladesh Bureau of Statistics.

————. 1994. *Population census of Bangladesh, 1991. Community Series (Meherpur)*. Dhaka: Bureau of Statistics.

————. 2001. *Zila Profile*. http://www.bangladeshgov.org/mop/ndb/zila-profile/meherpur.html. Accessed on 25 May 2001.

Grameen Bank. 2000. *Completion Report of Action Research on Community Based Arsenic Mitigation Project (A GB-DPHE-UNICEF project)*. Dhaka: Grameen Bank.

Greenberg, H. 1971. *Integer Programming*. New York/London: Academic.

Habib, A., et al. 2007. Effectiveness of arsenic mitigation program in Bangladesh—relationship between arsenic concentrations in well water and urine. *Osaka City Medical Journal* 53: 97–103.

Hassan, M.M. 2000. Approaches to the arsenic research in Bangladesh: A review. *Journal of the Bangladesh National Geographical Association* 27–28 (1–2): 45–59.

Hassan, M.M., and S.C. Das. 2000. Inorganic arsenic in Bangladesh: health and social crisis. *The Jahangirnagar Review, Part II: Social Science*, XXIII–XXIV, 139–155.

Hoque, B.A., et al. 2003. Appropriateness of arsenic mitigation water supply options: Bangladesh experience. *Environment, Sanitation, Engineering Research* 17 (3): 221–224.

————. 2006. Arsenic mitigation for water supply in Bangladesh: appropriate technological and policy perspectives. *Water Quality Research Journal of Canada* 41 (2): 226–234.

Hore, S.K., M. Rahman, M. Yunus, C.S. Das, S. Yeasmin, S.K.A. Ahmad, M. Sayed, A.M. Islam, M. Vahter, and L.A. Persson. 2007. Detecting arsenic-related skin lesions: Experiences from a large community-based survey in Bangladesh. *International Journal of Environmental Health Research* 17: 141–149.

Howard, G., and F. Ahmed. 2006. Identifying the preferred arsenic mitigation options in Bangladesh. *Journal of Health Population and Nutrition* 24 (3): 346–355.

Hussam, A., Ahamed S., and Munir A.K.M. 2008. Arsenic filters of groundwater in Bangladesh: Towards a sustainable solution. *National Academy of Engineering of the National Academies Publications*, 38(3). http://www.nae/bridgecom.nsf/MKEZ-7JLKJD?OpenDocument. Accessed 9 Jan 2009.

Iacofano, D.S. 1990. *Public involvement as an organizational development process: A proactive theory for environmental planning program management*. New York/London: Garland Publishing, Inc.

Inauen, J., and H.J. Mosler. 2016. Mechanisms of behavioural maintenance: Long-term effects of theory-based interventions to promote safe water consumption. *Psychology & Health* 31 (2): 166–183. doi:10.1080/08870446.2015.1085985.

Inauen, J., M.M. Hossain, R.B. Johnson, and H.J. Mosler. 2013. Acceptance and use of eight arsenic-safe drinking water options in Bangladesh. *PloS One* 8 (1): e53640. doi:10.1371/journal.pone.0053640.

Islam, K.Z., Md.S Islam, J.O. Lacoursière, and L. Dessborn. 2014. Low cost rainwater harvesting: An alternate solution to salinity affected coastal region of Bangladesh. *American Journal of Water Resources* 2 (6): 141–148. doi:10.12691/ajwr-2-6-2.

Jakariya, M. 2003. *The Use of Alternative Safe Water Options to Mitigate Arsenic Problem in Bangladesh: Community Perspective.* Dhaka: BRAC.

Jakariya, M., et al. 2003. Sustainable community based safe water options to mitigate the Bangladesh arsenic catastrophe—an experience from two upazillas. *Current Science* 85 (2): 14–146.

Johnson, B.L.C. 1975. *Bangladesh.* London: Heinemann Educational Books.

Kabir, A., and G. Howard. 2007. Sustainability of arsenic mitigation in Bangladesh: Results of a functionality survey. *International Journal of Environmental Health Research* 17 (3): 207–218.

Khan, M.M.H., et al. 2007a. Determinants of drinking arsenic-contaminated tubewell water in Bangladesh. *Health Policy and Planning.* 22 (5): 1–9.

Khan, M.M.H., A. Khandoker, M. Kabir, and M. Mori. 2007b. Determinants of drinking arsenic-contaminated tubewell water in Bangladesh. *Health Policy and Planning* 22 (5): 335–343. doi:10.1093/heapol/czm018.

Koopmans, T.C. 1951. *Activity Analysis of Production and Allocation,* Cowels Commission for Research in Economics, Monograph No. 13. New York: Wiley.

Kotler, P. 1975. *Marketing for Nonprofit Organizations.* Englewood Cliffs: Prentice Hall.

Kotler, P., and E.L. Roberto. 1989. *Social Marketing: Strategies for Changing Public Behavior.* New York/London: The Free Press.

Koundouri, P., K. Kemper, A. Talbi, and M. Sur. 2009. The economics of arsenic mitigation: investment choices—what and when? In *Arsenic Contamination in Groundwater in South and East Asian Countries: Towards a more Effective Response.* Vol. II, technical report of the World Bank, report no. 31303. http://siteresources.worldbank.org/INTSAREGTOPWATRES/Resources/ArsenicVolII_PaperIV.pdf. Accessed 13 Oct 2016.

Liu, F.F., C. Huang, and Y. Yen. 2000. Using DEA to obtain efficient solutions for multi-objective 0-1 linear programs. *European Journal of Operational Research* 126: 51–68.

Lockwood, B. 1987. Pareto efficiency. In *The New Palgrave: A Dictionary of Economics,* ed. J. Eatwell, M. Milgate, and P. Newman, vol. 3. London/New York/Tokyo: Macmillan Maruzen.

Lokuge, K., et al. 2004. The effect of arsenic mitigation interventions in disease burden in Bangladesh. *Environmental Health Perspectives* 112 (11): 1172.

Mallick, A.S.M.F., and H.K. Das. 2003. Options for arsenic removal from groundwater. In *Groundwater Resources and Development in Bangladesh: Background to the Arsenic Crisis, Agricultural Potential and the Environment,* 447–464. Dhaka: UPL.

Michael, H.A., and C.I. Voss. 2008. Evaluation of the sustainability of deep groundwater as an arsenic-safe resource in the Bengal basin. *Proceedings of the National Academy of Sciences of the United States of America* 105 (25): 8531–8536.

Milton, A.H., Z. Hasan, A. Rahman, and M. Rahman. 2001. Chronic arsenic poisoning and respiratory effects in Bangladesh. *Journal of Occupational Health* 43: 136–140.

Milton, A., W. Smith, K. Dear, J. Ng, M. Sim, G. Ranmuthugala, G. Lokuge, B. Caldwell, H. Raman, H. Rahman, A. Shraim, D. Huange, and M. Abrar. 2006. A randomised intervention trial to assess two arsenic mitigation options in Bangladesh. *Epidemiology* 17 (Suppl): S219.

Nickson, R., J. McArthur, W. Ahmed Burgess, P. Ravenscroft, and M. Rahman. 1998a. Arsenic poisoning of Bangladesh Groundwater. *Nature* 395: 338.

Opar, Alisa, et al. 2007. Responses of 6500 households to arsenic mitigation in Araihazar, Bangladesh. *Health & Place* 13 (1): 164–172. doi:10.1016/j.healthplace.2005.11.004.

Parsons, T. 1991. *The social system,* Newth ed London: Routledge.

Paul, B.K. 2004. Arsenic contamination awareness among the rural residents in Bangladesh. *Social Science and Medicine* 59: 1741–1755.

Philip, J. 1972. Algorithms for the vector maximization problem. *Mathematical Programming* 2: 207–229.

Prothom Alo. 2000, December 14. Bangla daily newspaper.

———. 2001, February 24. Bangla daily newspaper. "Solution to the Arsenic Problem." Special supplement.

———. 2017, April 21. Bangla daily newspaper. "Arsenic Situation in the Country" (*Deshe Arsenic Poristhiti*).

Quamruzzaman, Q.S., M.R. Roy, S. Mia, and A.I. Arif. 1999. Rapid action programme: emergency arsenic mitigation programme in two hundred villages in Bangladesh. In *Arsenic Exposure and Health Effects*, ed. W.R. Chappell, C.O. Abernathy, and R.L. Calderon, 363–366. Oxford: Elsevier.

Rahman, M., and O. Axelson. 2001. Arsenic ingestion and health effects in Bangladesh: Epidemiological observations. In *Arsenic Exposure and Health Effects*, ed. W.R. Chappell, C.O. Abernathy, and R.L. Calderon, 193–199. Oxford: Elsevier.

Rahman, M.M., et al. 2002. Effectiveness of reliability of arsenic field testing kits. *Environmental Science and Technology* 36 (24): 5385–5394.

Rahman, M.M., and F. Jahra. 2009. *Challenges for implementation of rain water harvesting project in arsenic affected areas of Bangladesh*. http://wepadb.net/pdf/0612sympo/paper/Md.MafizurRahman.pdf. Accessed on 12 Feb 2009.

Rammelt, C., M. Zahed, J. Boes, and F. Masud. 2011. Beyond medical treatment, arsenic poisoning in rural Bangladesh. *Social Medicine* 6 (1): 22–28.

Read, H. 1975. *Communication: Methods for all Media*. Urbana/Chicago/London: University of Illinois Press.

Sarkar, M.Q., M. Jakariya, and M. Rahman. 2005. *Assessment of the Rural Piped Water System in Arsenic Affected Rural Bangladesh*. BRAC research report, Dhaka, Bangladesh.

Schramm, W. 1971. The nature of communication between humans. In *The Process and Effects of Mass Communication* (pp. 3-51). Revised edition, W. Schramm and D. F. Roberts (eds.), Urbana/Chicago/London: Illinois Press.

Sewell, W.R.D., and J.T. Coppock. 1977. A perspective on public participation in planning. In *Public Participation in Planning*, ed. W.R.D. Sewell and J.T. Coppock. London: Wiley.

Shankar, S., S. Shanker, and Shikha. 2014. Arsenic Contamination of Groundwater: A Review of Sources, Prevalence, Health Risks, and Strategies for Mitigation. *The Scientific World Journal* 2014 (2014): 304524. doi:http://dx.doi.org/10.1155/2014/304524. Accessed 17 Feb 2017 at https://www.hindawi.com/journals/tswj/2014/304524/

Sierksma, G. 1996. *Linear and Integer Programming: Theory and Practice*. New York: Marcel Dekker, Inc..

Sikder, A.M., and M.H. Khan. 2002. *Mitigation of Arsenic Contamination in Ground Water in Rural Setting: An Action Research*, Gyan Mela publication series. Dhaka: Grameen Trust.

Smith, A.H., M.L. Biggs, L. Moore, R. Haque, C. Steinmaus, J. Chung, et al. 1999. Cancer risks from arsenic in drinking water. In *Arsenic Exposure and Health Effects*, ed. W.R. Chappell, C.O. Abernathy, and R.L. Calderon, 191–200. Amsterdam: Elsevier.

Smith, A., E.O. Lingas, and M. Rahman. 2000. Contamination of drinking-water by arsenic in Bangladesh: A public health emergency. *Bulletin of the World Health Organization* 78(9): 1093–1103.

Takayama, A. 1974. *Mathematical Economics*. Illinois: The Dryden Press.

Walukiewicz, S. 1991. *Integer Programming*. Dordrecht/Boston/London: Kluer Academic Publishers.

Yang, Jian-Bo. 2000. Minimax reference point approach and its application for multi-objective optimization. *European Journal of Operational Research* 126: 541–556.

Yokata, H.K., M. Tanabe, Y. Sezaki, T. Akiyoshi, K. Miyata, S. Kawahara, et al. 2001. Arsenic contamination of ground and pond water and water. *Geology* 60: 323–331.

Zeleny, M. 1974. *Linear multi-objective programming*. Berlin/Heidelberg/New York: Springer Verlag.

Zuberi, M.I. 2003. Arsenic mitigation through AMP-christian aid support: Awareness, communication and scaling-up: An assessment report commissioned by AMP-Christian Aid (UK).

Unpublished Materials

Adhikarya, R. 1989. Strategic extension campaigns on rat control in Bangladesh. In R. E. Rice, C. K. Atkin (eds.), Public Communication Campaign (pp. 230-232). Second edition, Beverly Hills; London; New Delhi: Sage Publications.

Ahmed, M. 2000. *Arsenic Problem in Bangladesh: An Overview, Resource Paper on Contamination/ Mitigation in Bangladesh for World Bank Policy Formulation* (submitted to the World Bank in April, 2000; paper procured from the author through personal communication).

Ahmad, N. 1976. Economic Geography of Bangladesh. New Delhi: Vikas Publishing House.

Ahmed, M.F. 2002. *Alternative Water Supply Options for Arsenic Affected Areas in Bangladesh.* Paper presented at the international Workshop on arsenic mitigation in Bangladesh, During 14–16 January, 2002, organized by The Local Government Division, Ministry of Local Government Rural Development & Cooperatives, Government of the People's Republic of Bangladesh at Dhaka: Centre for water supply and waste management.

––––––. 2004. *Cost of Water Supply Options for Arsenic Mitigation.* Paper presented at the 30th WEDC international conference, held in Vientiane, Lao, PDR in 2004.

BAMWSP. 2002. *Home page of Bangladesh Arsenic Mitigation and Water Supply Project.* http:// www.BAMWSP.org. Accessed 3 Mar 2002.

Barkat, A., and A. Hussam. 2008. Provisioning of arsenic-free water in Bangladesh: A human rights challenge. *Prepared for* Session I: Engineering and Special Vulnerabilities, at Workshop on *Engineering, Social Justice, and Sustainable Community Development.***Organized by** The National Academy of Engineering (NAE), Centre for Engineering, Ethics, and Society, (co-sponsored by the Association for Practical and Professional Ethics, and the National Science Foundation). Washington, DC: October 2–3, 2008.

Bridge, T.E., and M.T. Husain. 2000. *The Increased Draw Down and Recharge in Groundwater Aquifers and Their Relationship to the Arsenic Problem in Bangladesh.* http://www.eng-consult.com/arsenic/article/meerarticle6.html. Accessed 13 Oct 2016.

British Geological Survey. 2009. *Implications of the Present Study for the Arsenic Mitigation Strategy.*http://www.bgs.ac.uk/arsenic/bphase1/B_impl.htm. Accessed 13 Jan 2009.

Burren, M. 2000. *Small scale variability of arsenic in ground water in the district of Meherpur,* Western Bangladesh. Thesis submitted in partial fulfillment of Hydrogeology M.Sc., University College, London, UK.

Chowdhury, S. N. 1998. Safe drinking water for all: Containing arsenic contamination: an NGO initiative. The Daily Star 16 July, 1998.

Curry, A., Carrin, G., Barteam, J., Yamamura, S., Heijnen, J.S., Hueb, J., Sato, Y. (2000). Towards an Assessment of the Socioeconomic Impact of Arsenic Poisoning in Bangladesh. A monograph prepared for the World Health Organization (Department of Health in Sustainable Development and Water, Sanitation and Health), Geneva.

Department of Public Health Engineering (DPHE), Government of the People's Republic of Bangladesh and British Geological Survey (BGS), UK. 2000. *Groundwater Studies of Arsenic Contamination in Bangladesh: Final Report Summary.* Internet communication, June 25, 2000 at http://www.bgs.ac.uk/arsenic/Bangladesh/home.html.

––––––. 2001. *Groundwater Studies of Arsenic Contamination in Bangladesh: Final Report Summary.* http://www.bgs.ac.uk/arsenic/Bangladesh/home.html. Accessed 30 Dec 2015.

Dhaka Community Hospital. 2002. *Arsenic Problem.* http://www//dchtrust.org. Accessed 12 Apr 2002.

Dhaka Community Hospital (DCH) and School of Environmental Studies, Jadavpur University (SOES, JU). 2000. *Groundwater Arsenic Contamination in Bangladesh: (A) Summary of 239 Days Field Survey from August 1995 to February, 2000; (B) Twenty Seven days Detailed Field Survey Information from April 1999 to February 2000.* Unpublished survey report procured from Dr. Dipankar Chakraborty, Head of SOES, JU.

Edmunds, W.M., K.M. Ahmed, and P.G. Whitehead. 2015. A review of arsenic and its impacts in groundwater of the Ganges–Brahmaputra–Meghna delta, Bangladesh. *Environmental Science: Processes & Impacts.* doi: 10.1039/c4em00673a. http://www.espa.ac.uk/results/publications/review-arsenic-and-its-impacts-groundwater-ganges-brahmaputra-meghna-delta. Accessed 26 Aug 2015.

Flanagan, S. V., R. B Johnston, and Y. Zheng. 2012. Arsenic in tube well water in Bangladesh: Health and economic impacts and implications for arsenic mitigation. *Bulletin of the World Health Organization* 9, 839–846. doi: 10.2471/BLT.11.101253. http://www.who.int/bulletin/volumes/90/11/11-101253/en/. Accessed 22 Aug 2015.

Gadgil, A.J., and E.A. Derby. 2003. *Providing Safe Drinking Water to 1.2 Billion Unserved People. Submitted for Presentation to 96th Annual AWMA Conference*, San Diego, CA June 22–26, 2003.

Government of the People's Republic of Bangladesh. 2001a. *Zila profile*. http://www.bangladesh-gov.org/mop/ndb/zila-profile/meherpur.html. Accessed on 25 May 2001.

———. 2001b. *Demographic data*. http://www.bangladeshgov.org/mop/ndb/data-sheet/demo-data.html. Accessed 25 May 2001.

———. 2004. *National Policy for Arsenic Mitigation 2004*. http://www.dphe.gov.bd/pdf/National-Policy-for Arsenic-Mitigation-2004.pdf. Accessed 13 Jan 2009.

Hanchett, S., and T.H. Monju. 2009. The *Bangladesh Arsenic Mitigation and Water Supply Project: A Public Administration and Public Health Failure*. A paper presented in the session on water, culture, and power; American anthropological association annual meetings Philadelphia, Pennsylvania December 5, 2009.

Haque, A.K. Enamul, M.Z.H. Khan, and J Roy (Economic Research Group, Dhaka). 2009. *Impact of Arsenic Contamination in Groundwater on Poverty and Choice of Mitigation Technology for Rural Communities in Bangladesh*. http://www.saneinetwork.net/research/mir/2pdf. Accessed 12 Feb 2009.

Human Rights Watch. 2016. *Nepotism and Neglect: The Failing Response to Arsenic in the Drinking Water of Bangladesh's Rural Poor*. USA: Human Rights Watch. https://www.hrw.org/sites/default/files/report_pdf/bangladesh0416web_0.pdf. Accessed 13 Oct 2016.

Hussam, A., S. Ahamed, and A.K.M Munir. 2008. Arsenic filters of groundwater in Bangladesh: Towards a sustainable solution. *National Academy of Engineering of the National Academies Publications*, 38(3). http://www.nae/bridgecom.nsf/MKEZ-7JLKJD?OpenDocument. Accessed 9 Jan 2009.

HVR Arsenic Project. 2000. *What is Arsenic?*http://www.hvr.se/nov97/arsenic.html. Accessed 23 May 2000.

Jakariya, Md. 2000. *The Use of Alternative Safe Water Options to Mitigate the Arsenic Problem in Bangladesh: A Community Perspective*. M.Sc. thesis, Department of Geography, University of Cambridge, U.K.

Jones, E.M. 2000. *Arsenic 2000: An Overview of the Arsenic Issue in Bangladesh*. Paper presented at the 32nd WEDC International Conference, Colombo, Sri Lanka. http://phys4.harvard.edu/~wilson/wateraid.pdf. Accessed on 2 June 2001.

Kabir, M.A., and R. Johnston. 2007. *Performance of Arsenic-Removal Filters Used in Rural Bangladesh*. Paper presented at the 11th annual scientific conference, 2007 held at ICDDR,B, Dhaka, Bangladesh during March 4–6 March, 2007.

Kamruzzaman, A.K.M., and F. Ahmed. 2006. *Study of Performance of Existing Pond Sand Filters in Different Parts of Bangladesh*. Paper presented at the 32nd WEDC international conference, Colombo, Sri Lanka.

Kawahara, K. 2000. *Mitigation of Arsenic Contamination in Rural Areas of Bangladesh*. http://www.asia-arsenic.net/dhaka/kk-repole.html. Accessed 31 July 2001.

Koundouri, P., K. Kemper, A. Talbi, and M. Sur. 2009. The economics of arsenic mitigation. In *Arsenic Contamination in Groundwater in South and East Asian Countries: Towards a More Effective Response*. Vol. II, technical report of the world bank, report no. 31303. http://siteresources.worldbank.org/INTSAREGTOPWATRES/REsources/ArsenicVol.II_PaperIV.pdf. Accessed 12 Feb 2009.

Krajick, Kevin., and D. Funkhouser. 2015. *Battling 'the Largest Mass Poisoning in History'*. http://www.rawscience.tv/battling-the-largest-mass-poisoning-in-history-arsenic-in-bangladesh/. Accessed 22 Aug 2015.

Kunikane, Shoichi. 2009. *An Integrated Approach to Arsenic Mitigation in Bangladesh*. http://www.who.int/household_water/resources/kunikane.pdf. Power-point presentation Accessed on 7 Jan 2009.

Mortoza, S. 2000. Arsenic revisited. http://www.dainichi-consul.co.jp/english/arsenic/article/sylvia1.html. Accessed on 23 Apr 2000.

Murcott, Susan. 1999. *Appropriate Remediation Technologies for Arsenic Contaminated Wells in Bangladesh*. A paper presented at international conference on arsenic on Bangladesh ground water: The World's greatest arsenic calamity-- held at Wagner College, New York, U.S.A. Feb. 27–28, 1999. http://phys.harvard.edu/~wilson/murcott.html Accessed 20 July 2000.

Nahar, B.S. 2000. *Water Quality and Gender Inequality: Arsenic Contamination of Bangladesh's Groundwater and the Impact on the Lives of Rural Women*. Research monograph for DANIDA Bangladesh.

Nahar, N., and T. Honda. 2006. *Arsenic Mitigation Technologies in Bangladesh: Evidence from the Literature*. http://www.bdiusa.org/Journal%20of%20Bangladesh%20Studies/Volume%208.2%20(2006)/ARSENIC%20MITIGATION%20TECHNOLOGIES%20IN%20BANGLADESH.pdf. Accessed 26 Aug 2014.

Prothom Alo. 2001. *Solution to the arsenic problem*. Special supplement. February 24, 2001. Bangla daily newspaper.

Prothom Alo 2017. Bangla daily newspaper. Arsenic Situation in the Country (Deshe Arsenic Poristhiti). April 21

Rahman, M.M., and F. Jahra. 2015. *Challenges for Implementation of Rain Water Harvesting Project in Arsenic Affected Areas of Bangladesh*. http://wepa-db.net/pdf/0612sympo/paper/Md.MafizurRahman.pdf. Accessed 30 Dec 2015.

Rondinelli, D.A. 1977. *Planning development projects*. Stroudsburg: Dowden, Hutchinson & Ross Inc.

Sarkar, Prafulla. C. 1999. *Beliefs and Arsenicosis and Their Impact on Social Disintegration in Bangladesh: Challenges to Social Work Intervention*. A paper presented at the 26th A. A. S. W. national conference in conjunction with the Asia Pacific regional conference of the I. F. S. W. and A. P. A. S. W. E., September 26–29, 1999.

Sutherland, D., M.O. Kabir, and N.A. Chowdhury. 2009. *Rapid assessment of technologies for arsenic removal at the household level*. http://www.unu.edu/env/Arsenic/Sutherland.pdf. Accessed 12 Feb 2009.

Uddin, M.S., and A.S Khan. 2004. Water quality and follow-up survey on arsenic contamination of dugwells in Sharsha Upazilla. JICA/AAN Arsenic Mitigation Project Report 2.

———. 2004. *Arsenic Contamination of Deep Tubewells in Sharsha Upazilla*. http://project.jica.go.jp/bangladesh/0510532E0/09/pdf/01.pdf. Accessed 21 Jan 2009.

UNICEF, Bangladesh. 1999. *Arsenic mitigation in Bangladesh, media brief*. http://www.unicef.org/arsenic/. Accessed 15 Feb 2001.

Virtual Bangladesh. 2001. *The grand tour (Economy)*. http://www.virtualbangladesh.com/bd-econ-facts.html. Accessed 27 May 2001.

Wagelin, M. et al. 2000. *SODIS-An Arsenic Mitigation Option?* Paper presented at the 26th WEDC conference (Water, sanitation and hygiene: Challenges of the Millennium), held in Dhaka, Bangladesh, 2000.

Waterhealth International Inc. 2001. Products. http://www.waterhealth.com.cws.html. Accessed 27 May 2001.

Website of Mujibnagar. 2015. http://www.mujibnagar.com/mujibnagar/meherpur-district Accessed 09 Oct 2015.

Website of the Department of Public Health Engineering, Bangladesh. 2016. http://www.dphe.gov.bd/index.php?option=com_content&view=article&id=96&Itemid=104. Accessed 16 Oct 2016.

World Bank. 2007. Implementation completion and results report (ida-31,240 swtz-21,082) on a credit in the amount of SDR 24.2 Million (US$ 44.4 Million Equivalent) to Bangladesh for arsenic mitigation water supply report No: ICR000028 June 10, 2007. Sustainable development department environment and water resources unit, South Asia Region.

————. 2011. Implementation, completion and results report (Ida-H1010) on a grant in the amount of SDR 27.6 Million (Us$ 40.0 Million Equivalent) subsequently reduced after restructuring to the amount of SDR 11.8 Million (Us$ 18.6 Million Equivalent) to The people's republic of Bangladesh for a water supply program project. Report No: ICR507 June 14, 2011. Urban, Water and disaster management unit sustainable development department South Asia region.

World Health Organization. 2000. *Towards an Assessment of the Socioeconomic Impact of Arsenic Poisoning in Bangladesh.* http://www.who.int/water_sanitation_health/dwq/arsenic2/en/print. html. Accessed 13 Jan 2009.